THE MORAL PSYCHOLOGY OF INTERNAL CONFLICT

This book pushes back against the potential trivialization of moral psychology that would reduce it to emotional preferences and takes an enactivist, self-organizational, and hermeneutic approach to internal conflict between a basic exploratory drive motivating the search for actual truth, and opposing incentives to confabulate in the interest of conformity, authoritarianism, and cognitive dissonance, which often can lead to harmful worldviews. The result is a new possibility that ethical beliefs can have truth value and are not merely a result of ephemeral altruistic or cooperative feelings. It will interest moral and political psychologists, philosophers, social scientists, and all who are concerned with inner emotional conflicts driving ethical thinking beyond mere emotivism, and toward moral realism, albeit a fallibilist one requiring continual rethinking and self-reflection. It combines "basic emotion" theories (e.g., Panksepp) with hermeneutic depth psychology. The result is a realist approach to moral thinking, emphasizing coherence rather than foundationalist theory of knowledge.

RALPH D. ELLIS received his PhD from Duquesne University and a postdoctoral MS from Georgia State University. A lifelong practitioner of Gendlin's "focusing" method, he has worked as a social worker as well as teacher, and is interested in integrating the social sciences with enactive consciousness theory. His books include *An Ontology of Consciousness*, *Theories of Criminal Justice*, *Coherence and Verification in Ethics*, *Questioning Consciousness*, *Eros in a Narcissistic Culture*, *Just Results: Ethical Foundations for Policy Analysis*, *Love and the Abyss*, and *Curious Emotions;* he has also co-authored a book with Natika Newton on enactivist consciousness theory called *How the Mind Uses the Brain.*

THE MORAL PSYCHOLOGY OF INTERNAL CONFLICT

Value, Meaning, and the Enactive Mind

RALPH D. ELLIS

Clark Atlanta University

CAMBRIDGE
UNIVERSITY PRESS

CAMBRIDGE
UNIVERSITY PRESS

University Printing House, Cambridge CB2 8BS, United Kingdom

One Liberty Plaza, 20th Floor, New York, NY 10006, USA

477 Williamstown Road, Port Melbourne, VIC 3207, Australia

314–321, 3rd Floor, Plot 3, Splendor Forum, Jasola District Centre, New Delhi 110025, India

79 Anson Road, #06-04/06, Singapore 079906

Cambridge University Press is part of the University of Cambridge.

It furthers the University's mission by disseminating knowledge in the pursuit of education, learning, and research at the highest international levels of excellence.

www.cambridge.org
Information on this title: www.cambridge.org/9781107189959
DOI: 10.1017/9781316996683

First published 2018

Printed in the United Kingdom by Clays, St Ives plc

A catalogue record for this publication is available from the British Library.

ISBN 978-1-107-18995-9 Hardback

To the memory of Gene Gendlin and Jaak Panksepp, who each profoundly influenced me in their own ways

Contents

Acknowledgments

For many years, I have been concerned that moral psychology may be driving us toward an increasing relativization of ethics. The need for moral behavior to be motivated has led to an emphasis on natural empathic emotions as the driving force for morality. What worries me is that, if we are to act morally when and if our empathic emotions happen to motivate us, then the general rule becomes the following: Always do whatever you want to do (benevolent if that is the way you happen to feel, and implicitly, malevolent if that is how you happen to feel). This book is my attempt to confront that problem after many years of studying enactivist and self-organizational approaches to consciousness and emotion.

The development of this project is almost completely thanks to the kindness of many psychologists and philosophers who shared their knowledge and insights with me. Jaak Panksepp generously discussed his research and ideas with me, from which I was able to get some insight not only into his findings, but also his own interpretations of them. Some of the most central arguments of this book developed through a series of conversations with Charlie Harvey, who steered me in better directions with regard to the phenomenological dimension. I was lucky enough to read Eugene Gendlin's *Experiencing and the Creation of Meaning* as a very young man – an exposure that dramatically influenced the entire direction of both my intellectual and personal development. In the area of enactivist cognitive theory, I would never have known without the help of Natika Newton how to navigate the challenging realm of self-organization and dynamical systems, especially as they connect to emotion. I was also helped, over a period of many years, by neuropsychologist Bill Faw, who has a unique ability to break down brain processes in a way that is remarkably intelligible to the novice that I still am in this area. The results of many exchanges with Peter Zachar in the realm of clinical psychology also show up in this book, especially around the issues of authoritarianism and conformity theory and the interplay of biology with environmental influences in developmental psychology. I was also helped immensely by collaborations with Louise Sundarajan, a gifted clinical psychologist.

Randy Auxier, a former editor of *The Pluralist*, has patiently argued through with me many of the crucial issues discussed here, and also has given me kind encouragement throughout most of my academic career. The organizers and participants of the Society for Philosophy in the Contemporary World have offered amazing forums for serious and constructive discussion; part of the argument of this book was put forward in a simpler form in a 2017 article in the Society's journal. I was also helped immensely by the Society for the Scientific Study of Consciousness. Anthony Freeman, one of the spearheads of that movement, especially encouraged me to become increasingly involved in that area.

Maxim Stamenov played a pivotal role in my life trajectory. As a person who had approached consciousness from both an empirical and phenomenological perspective, I had been virtually irrelevant to the English-speaking cognitive theory discussion, which was dominated by extreme reductionism at the time, until Stamenov read my early work *An Ontology of Consciousness* and invited me to write a volume for his *Advances in Consciousness Research* series, just at the time when the study of consciousness was entering what I believe was a Kuhnian revolutionary situation. Without that invitation, I might never have gotten involved with any of the above-mentioned people who helped me so much in the study of consciousness from empirical, phenomenological, physiological, and clinical perspectives. From the philosophical standpoint, I am grateful for the educating exchanges I have had with Andy Fiala, David Carr, Tom Nenon, David Chan, Nick Georgalis, John Bickle, Norm Fischer, and a much longer list. I am indebted to the editors of the *Cilicia Journal of Philosophy* for allowing me to use some passages from my 2016 article for them. Finally, I have to thank my graduate students at Clark Atlanta University for constantly challenging and pushing me on many of the issues discussed here.

PART I

Love of Truth and "Moral Sentiments"

CHAPTER 1

The Paradox of the Charitable Terrorist

At New York's Grand Central Station recently, a Muslim woman was physically attacked and injured for wearing a hijab, while bystanders watched and refrained from interfering (NBC New York 2016). Such daily news stories may sometimes prompt us to remark that people "ought" to be better than they are. But this complaint remains vacuous unless something in human nature is capable of motivating us to do what we "ought."

The human sciences, whether empirical, phenomenological, or "postmodern," often address this problem by focusing on altruistic and cooperative emotional tendencies that occur naturally – either innately or through socialization. I want to push back against this approach. Superficially, it seems "natural" to focus on such emotions. After all, action must be motivated. Moral psychologists standardly say that "ought implies can": There is no point in saying that someone who can't swim ought to save a drowning person. In the same way, an "ought" that can't be motivated is either meaningless or false.

The "why be moral?" problem (Nielsen 1989) is a variant of this same challenge. What is it in our human makeup that motivates us to *care* about the effects of our actions on other people? Is it merely a socially conditioned habit? Is it a mutually calculated cooperation for purposes of self-interest, as Hobbes believed? Is it a feeling of empathy toward specific individuals with whom we happen to identify? Or is there something more than that? Philosophers and psychologists for the past century or more have relied on "naturally altruistic emotions" such as nurturance and social bonding to answer this question. Given the survival value of cooperation, they theorize that humans have adapted by developing a natural desire to be helpful. In short, we ought to be moral because we naturally want to.

But this way of thinking about moral psychology can lead ethical thinking down a dangerous rabbit hole. If we can't do anything other than what our altruistic (or malevolent) emotions drive us to do, then ultimately the only moral guidance we can expect is "Do whatever you are already motivated to do."

In other words, "Do whatever you want." Such a moral system seems frustratingly vacuous, even pointless. At the end of the film *Baby, the Rain Must Fall*, we hear the song lyric "Wherever my heart leads me, that's where I must go" as the main character is being led away in handcuffs. Is there not a more promising path than "You should do X whenever you are naturally motivated to do X?" We might as well say "Be altruistic if and only if you feel like it, and be malevolent whenever you feel like it." Even a sociopath could cheerfully follow this injunction.

Luckily, naturalism doesn't need to bet its entire hand on the notoriously fickle "altruistic emotions." Contemporary emotion researchers like Jaak Panksepp (1998, 2000, 2011), Doug Watt (2000), and Nico Frijda (2006/2007) have noticed that the search for moral meaning, like any other everyday truth-seeking activity, is energized not only by empathic instincts or social conditioning, but also and more importantly it is grounded in a basic exploratory drive. An exploratory drive means that we want to know what the truth is about reality (and in principle this would include sociopolitical and ethical reality), *independently* of whether we are instrumentally rewarded for seeking it, and also independently of whether we happen to feel altruistic or nurturing in a particular instance (see especially Panksepp 1998, p. 145 and passim).

Panksepp connects this relatively independent exploratory drive to what he calls an innate "SEEKING" system in the brain (Panksepp uses the all-caps to indicate a term of art in his nomenclature of basic emotion systems in the brain). The SEEKING system includes (among other tendencies that we will get to later) a desire to explore, simply in order to discover the truth about our world. Although widely distributed in the brain, this system uses specific brain areas and specific combinations of neurotransmitters that are altogether different from those used by other emotional brain systems. The exploratory drive naturally leads to curiosity (see Ellis 1995, 2005) as well as a tendency to "appraise" environmental contexts even before any specific goal has been set (Frijda 2006/2007, esp. pp. 78ff; Salmela 2014). This innate, unconditioned motivational system can lead to what the earlier psychologist Robert White (1959) called a "drive toward mastery" – a non-derivative desire to understand the environment and "master" our own self-motivated actions relative to the world as we grasp it. An exploratory drive would motivate us to seek the truth, and yet – importantly for our purposes here – would not predetermine *what* we believe to be true.

The *independence* of this exploratory drive – the fact that it is not derivative from any other emotion system or social reinforcement – has a crucial philosophical implication. Adherence to ethical principles might extend further than merely behaviors driven by cooperative or benevolent emotions. Naturalistic ethical theorists may not need to relegate moral beliefs exclusively to the realm of emotional feelings at all, whether innate or learned. Such beliefs might actually have *truth value.* Our moral comportment might then be driven by the results of ethical thinking, even when such results run contrary to any combination of desires or survival instincts *other than* the demands of the exploratory drive itself.

It is true, of course, that some people, at least a great deal of the time, don't *experience* themselves as caring much what the truth is about moral issues. But what I want to argue is that in many cases people are thinking of "moral" issues as a narrower category than the more general question as to "what we ought to do" per se. What some people designate as "moral" values are thought of by others as "non-moral values." The only clear difference, I will argue, is that "moral" values involve the harm or well-being of other valuing creatures. Rather than thinking in terms of "moral" principles, some people regard such behaviors simply as "being civil" or "being helpful." Other times, people have already presupposed (for whatever reasons) that there is no truth regarding "moral" issues – or that there is no truth that they don't already know. But if they did think there was moral truth to be discovered, an independent exploratory drive implies that they would be curious, because the desire to explore the truth extends to all domains that inherently require our attention, except when suppressed by the conflicting emotions that we will also be discussing here.

To be sure, not everyone is equally interested in exploring all the *same* aspects of reality. Some are more interested in chemistry or physics than music or math or philosophy. But when an aspect of reality *necessarily demands* a judgment, we have a natural interest in knowing what is actually true; we don't just automatically believe whatever we want. And the realm of value and meaning, I will argue, inevitably does call for our attention. At the same time, I argue that our experiential *awareness* of this basic truth-seeking interest results from a net summing, beneath the surface, of interacting and equally powerful conflicting motives – just as the experience of green can result from a net summing of blue and yellow. The motive to explore the truth often does come into conflict with other emotional tendencies that are equally powerful.

Frijda uses the term "action readiness" (p. 26ff and passim) somewhat similarly to Panksepp's notion of innate emotional-behavioral dispositions, especially the SEEKING system. In Frijda's estimate, we appraise situations in terms of their action potential, relative to the action readiness that we *already have.* "The emotive states are not mere wishes. They are actual, embodied states, or states on the verge of embodiment in action, *to be released when circumstances permit*" (p. 27, italics added). These action readinesses, as they play out in complex interactions with specific real-life situations, trigger what Frijda calls "passions" to act in certain ways, sometimes to achieve a goal, but sometimes just for the sake of exercising the action pattern itself.

Some of these action patterns comprise exploration of reality. Frijda borrows Charlotte Bühler's term *"Funktionslust"* – roughly, actions that serve as their own reinforcement without the need for instrumental reward – to designate this type of action readiness (Frijda 2007, p. 78 and passim; see also Bühler 1931). This concept of *Funktionslust* is highly suggestive of Panksepp's SEEKING system – a physiological emotion system that energizes a desire to activate ourselves toward exploration of our world, not necessarily to seek pleasure maximization even indirectly as an outcome. Panksepp concludes that this kind of innate motivational system triggers emotions and behavior patterns independently of hedonistic reward. One result is that we are designed to seek the truth, even though the truth sometimes may not lead to any ulterior reward – at least, it may not reward the individual member of the species on that occasion. So, even though Frijda's "appraisal theory" differs in some ways from Panksepp's "basic emotion" approach, they are in remarkable agreement on the issues that are relevant for our purposes. Higher animals are motivated to act *upon* their environment, not just react *to* it.

The new research in emotional neuropsychology may actually somewhat vindicate those who resisted strict "reinforcement" theories during the twentieth century, including White as well as Otto Rank (1936/1978), Carl Rogers (1951, 1961), Rollo May (1950/2011), K. G. Montgomery (1954), J. J. Gibson (1986), and Eugene Gendlin (1962/1997, 1978/1982). Recent naturalistic understandings of the brain tend to back up those departures from exclusively hedonistic reductionism – especially if by "hedonistic reinforcement" we mean instrumental rewards *external to the behavior itself.*[1] SEEKING behavior is its own reinforcement. The desire to know the truth is not merely a conditioned skill set in the service of other emotions; it is an innately grounded motivation in its own right. Panksepp, who has run several research labs studying the emotional brain and the affective behavior of mammals (including neurological studies of humans), explicitly repudiates explanations based entirely on reinforcement theories such as behaviorism (1998, pp. 147–150). In the final analysis, Gendlin (1978/1982) suggests that the motivation to explore both our affective qualities and their environmental referents is a difficult, nuanced process, and can't be entirely understood in terms of instrumental hedonistic outcomes.

Because some motivations are independent of external rewards, we will also find as we proceed that the SEEKING system not only motivates the search for truth. It also grounds a continually present affective quality of "zest for life," which in turn enables everyday feelings of "inspiration" in a mundane but crucial sense. Panksepp et al. (2014) speak of clinical depression as the absence of a feeling of "enthusiasm," which is rooted in the SEEKING system. (The evidence for this view includes brain scans and chemical manipulation of neurotransmitters in various experimental and clinical settings, as summarized extensively by Panksepp 1998, Panksepp et al. 2014, and other sources.) This feeling of wanting to act and to live is necessary to avoid clinical depression, independently of hedonistic considerations. Depression is not just sadness, sour disposition, or "dysthymic affect" – the "ain't it awful" feeling that we all have from time to time, which can be quite normal and functional. More than that, it includes a lack of motivation to take action as opposed to mere reaction. The depressed person can't feel the value of any achievable outcome strongly enough to motivate taking an initiative.

So the same emotion system that creates curiosity and the desire to explore reality also grounds the *intensity* with which we *value* any potential actions we might take, or their potential results. Because the intensity of valuing is a function of inspiration, I will argue that the SEEKING system also fuels the inspiration to try to act "in accord" with any values we believe we have discovered – and again, this would include moral values. But paradoxically, the *questioning* of why we take what we take to have value is also part of the exploratory drive, which in turn is another part of the same innate and independent SEEKING system. The same system that undergirds questioning also enables the feeling of inspiration.

The term "value" can be used in different senses. One of the important senses that concern us here is the one that is derivative from *but not reducible to* a subjective experience of valuing something. "X has value," when used in an objective sense, can be taken to mean that any rational being with a capacity for empathy *would* value X if attention were directed to X. In this sense, X usually has to do with the value of sentient beings and their well- or ill-being. This issue will be unpacked more precisely in Chapter 4.

It seems counterintuitive yet true that the same SEEKING system whose dampening is most likely to result in *depression* is precisely a system that functions relatively independently of other systems, including those involving *hedonistic reward.* On a purely hedonistic model, one might expect that eating enough potato chips should alleviate the sadness of depression, but it fails to do so, ultimately because the pleasure of the potato chips mostly affects a different brain system that is largely irrelevant to the depression – a separate PLEASURE/PAIN system, as we will discuss in the next chapter.

In Panksepp's assessment, what defines the SEEKING system as a *separate* system is that it doesn't need to be instrumentally rewarded in order for the relevant exploratory and attention-directing behaviors to be elicited. On the contrary, certain self-motivated behaviors, including exploration, are consistently orchestrated by endogenous and unconditioned combinations of brain areas and neurotransmitters, distinct from and not derivative from other emotional brain systems, such as the one that Panksepp refers to as the PLEASURE/PAIN system. We know this because of neurological studies involving deficits of one or another brain system, as well as by observing the action of neurotransmitters in the various emotion systems (for example, Frijda 2006; Panksepp 1998; Ellis 2005; some behavioral studies supported this idea of an independent exploratory drive as far back as Montgomery 1954; Harlow 1950; and Kagan and Berkun 1954).

On the other hand, even if there is a natural exploratory drive, it often would come into conflict with equally powerful incentives to confabulate, or simply to wear blinders to block out inconvenient or potentially disturbing truths, especially in the face of fear and anxiety. Both empirical and experiential analyses bear out this common-sensical notion of "motivated selective attention." We often temporarily abandon our truth-seeking tendency for various emotional reasons.

In fact, when we act immorally or subscribe to harmful ethical and social viewpoints, the new trends in emotion research call into question the common assumption that such moral lapses result simply from a deficit of empathy or "fellow feeling." The problem instead may often stem from a *selective suppression of the exploratory drive.*

Specifically, the exploratory drive can be selectively suppressed in certain quite circumscribed contexts when applied to the intellectual contemplation of moral and social questions (if I may use "intellectual" in a general and non-grandiose way). The proponents of Arian superiority, however intelligent or educated they may be, fail to ask themselves some of the most critical questions about the factual and logical basis of their theory – questions that seem obvious to others.

Hence the paradox of the Ku Klux Klan terrorist who conspires to assassinate civil rights leaders on Saturday, and then gives generous charitable donations to his church on Sunday morning, perhaps also volunteering at a local soup kitchen in the afternoon. The KKK Grand Wizard of Georgia (see Federal Bureau of Investigation 1964), for many decades, owned and operated Stone Mountain Park purely as a generous public service, free of charge and at considerable expense to himself (New York Times 1993). The Klan, the White Citizens Councils, and other such racist organizations were supported by charitable and public spirited leaders of many Southern US communities.[2] Many of them served on the committees of charities like the American Legion, the Community Chest, and the Legion's Honor Society (Luce 1960, p. 35ff). Many routinely tithed 10 percent of their income to their churches and organized programs to help the poor or the hungry, sometimes helping even poor African-American families. Obviously, these otherwise good people had no shortage of empathy and compassion – except when their entire philosophical worldview was threatened. On my view, that is when the exploratory drive is most quickly and *selectively* suppressed, and thus critical thinking – with corresponding selectivity – is severely diminished with respect to the particular philosophical issue at stake, as well as all logically interconnected beliefs and attitudes. The suppression of the exploratory drive, when it occurs, tends to be sculpted in bas relief against an otherwise intelligent background – more with a scalpel than with a butcher knife. The charitable person usually becomes a terrorist only in specific circumstances – most often a situation in which an entire moral worldview is threatened.

Such moral lapses can't be explained merely in terms of conformism, as highlighted in studies by Duriez and Soenens (2009), among others. Duriez and Soenens, working in Belgium, show that sheer conformism accounts for only a small part of the intercorrelations between "Authoritarian Personality," "Social Dominance Orientation" (SDO), and "Racism" (each measured by a separate questionnaire). Instead, the *extent* to which racism, SDO, and Authoritarianism Personality correlate depends largely on the extent to which Authoritarian Personality per se is passed down from parents to children – as measured by the intergenerational correlations with regard to Authoritarian Personality. Similarly, Carl Bell (1978, 1980) finds only very modest connections between conformism and racism, but with a much stronger connection to narcissistic disturbance.

So in our example of the White Citizens Councils, something *prior to* conformity had to explain why the upstanding citizens who formed the Councils would choose to consort with the seedier elements of white supremacist culture in the first place. Why choose *that* particular group to conform to? Beyond mere conformism, Duriez and Soenens suggest that certain structures of *belief formation* are favored by certain people in certain specific contexts. Irrational tendencies such as racism and authoritarianism spring from belief systems that correlate with these particular patterns of thinking. Only after a significant mass of people have *already* developed a certain way of thinking can others then adopt it for merely conformist reasons.

It is true that people sometimes act in harmful and cruel ways just because they want to follow the orders of an authority figure, as in the famous Stanley Milgram experiments (for example, Milgram 1974). But we will see as we proceed that this authoritarian tendency itself is rooted in a momentary selective suppression of the exploratory drive, which triggers some people (by no means all, nor in all circumstances) to refrain from questioning the authority's commands. In the Milgram studies, subjects would naturally have questioned the authority's command to deliver an "extremely hazardous" electrical shock possibly leading to "severe bodily injury," had not the exploratory drive been suppressed in that instant. Beyond the mere observation of simple authoritarianism, we need to understand why people sometimes succumb to this tendency, and other times don't.

In fact, some replications of the Milgram experiments fail to correlate willingness to deliver the shock with measures of authoritarian personality at all. Instead, the subject positively empathizes with the experimentor more than with the other "subject" who is in a different room, and therefore elicits less empathy (Bräten 2013, p. 158ff). To be sure, some people are more susceptible to this random manipulation of their empathy than others, but I want to argue that the explanation requires more than understanding a lack of empathy per se; it requires delving into the person's overall set of moral principles and the seriousness with which they have been thought through.

Even the "negativity bias" hypothesis – that a tendency to hyperactivate the brain's *fear* circuits correlates with racism and other such irrational moral sentiments (Hibbing et al. 2013) – is only a small part of the story. It fails to account for the phenomenon of the charitable KKK terrorist, whose negativity bias is activated only in highly selective circumstances. Many racists can also be quite open and friendly, as evidenced by the failure to significantly correlate racism with a lack of "agreeableness" as measured by the Big Five personality inventory (Hodson et al. 2009). In fact, in the Hodson et al. study, "agreeableness" actually correlated *positively* with "Right Wing Authoritarianism." The specific times when sentiments like racism, prejudice, and generally harmful philosophical attitudes are *selectively* brought out, I will argue, are the times when a person's entire presupposed worldview is in danger of being shaken.

Some might say that racism and xenophobia stem simply from "clannism," which is just a natural human tendency – that people "naturally" are hostile toward those who are outsiders to the group, and therefore tend to see the outsiders as valueless. But this everyday assumption is not actually supported by anthropological evidence, which shows a good deal of fluidity among primitive tribes (for example, see Langness 1977). In fact, people migrated frequently from one tribe to another, and daughters were required by the tradition of exogamy to marry outside of their clan. Many traditional cultures included a moral principle of hospitality toward outsiders, as well as intertribal games similar to our modern Olympic games. The natural need to be cautious toward strangers seldom led to mutual hostility until agriculture, land ownership, and overshooting of population niches became prominent about 15,000–20,000 years ago (Olson 2002). Skeletal remains prior to that point show little evidence of wounds from warfare, despite the occasional murder (sometimes resulting from "vendetta" retaliation as a necessary means of law enforcement). These observations don't entail a "noble savage" notion, but simply that early humans suffered from the same internal conflicts that we all do.

What I am proposing here is that exaggerated clannism, when it does occur, results like many other moral tensions from an internal conflict between two equally natural tendencies. On the one hand, the exploratory drive motivates us to think about value issues (including moral ones) and ask ourselves whether the truth implies that all valuing creatures deserve our consideration – at least prima facie, or "all else being equal." Even the welfare of animals apparently was taken into account by early humans, who are generally believed to have said prayers of apology to animals that had to be killed for food (for example, see Graham Harvey 2005).

The opposing tendency is more complex, resulting in a real-time partial blockage of the natural human desire to think coherently about moral issues – not the *ability*, but rather the *desire* to do so. This more complex selective blockage of the exploratory drive, just at the specific point when exploration begins to threaten prior moral assumptions, calls for careful study. In the process, we should always remember that the brains of the earliest homo sapiens were genetically almost exactly the same as our brains are today (granted a tiny admixture in modern humans of Neanderthal and other early human species). So the facile notion that intelligent, thoughtful systems of morality are a development of "modern civilizations" must always be questioned. I will argue on the contrary that moral thinking is largely a natural aspect of the development of the human brain through childhood and adolescence, although this natural moral development through brain growth can be somewhat tweaked by environmental experience. Peter Zachar (2000a) explores from a clinical psychology perspective the way these biological factors, even though rooted in brain function, can also be socially tweaked, or even influenced by "talking therapies." Even natural biological brain growth shows a good bit of situational "plasticity."

I am not suggesting that the suppression of the exploratory drive is an all-or-nothing, or even a personality trait. On the contrary, we all suppress it to varying degrees in various circumstances. The fact that we all sometimes *selectively* suppress it is partly why social and political questions are so interesting, and why critical dialogue in this realm is so indispensable. Our own vulnerability to such foibles is also a reason why delving seriously into moral psychology can be disturbing or even threatening. We may discover that some of our own unconscious presuppositions are rooted in quicksand. At the point when this threat erupts, the questioning function of the prefrontal brain areas (described by Luria 1980, Posner 1990, and others) suddenly becomes deliberately quiet. In everyday, experiential terms, we momentarily go into a partial "brain freeze," and then try to recover from the feeling of confusion by simply returning to our previous commitments. We will explore later some studies indicating that this type of momentary "brain freeze" is often reflected in the mutual inhibition of different brain systems when they come into conflict with each other.

The Klansman's selective suppression of the exploratory drive – thus a dampening of his desire to seek the truth in certain specific contexts even though he might be an excellent rocket scientist or auto mechanic – might be expected to extend beyond questions about individual conduct in narrowly circumscribed situations. Threats to the coherence of his overall worldview would necessarily extend also to a defensive posture regarding more general beliefs that go into a person's entire *interconnected* understanding of reality as a whole. In the attempt to maintain coherence with a previously formed worldview, this selective exploratory-drive shutdown could include broader social, political, and even religious elements. The Bible and even the theory of evolution must be interpreted in ways consistent with a theory that rationalizes racial domination, for example. Once such a convoluted belief system has taken shape, complete with all the ad hoc hypotheses needed to prop it up, the same process could then indirectly influence views about purely scientific topics such as global warming or the biology of homosexuality – anything that threatens the coherence of the person's already existing overall understanding of the world. Empirical psychology studies show statistically well-controlled correlations among many of these interrelated social, political, theological, ethical, and even scientific attitudes and beliefs, *independently* of a person's education or general intelligence (Adorno et al. 1964; Altemeyer 2008; Bell 1978, 1980; Duriez and Soenens 2009). Those who question the robustness and reliability of these correlations should take a quick look at *The Authoritarian Personality* by Adorno et al. (1964), whose laborious statistical analyses could also be used as a test of the attention span of even the most avid statistics buff.

1.1 Empathy and the Exploratory Drive in Moral Development

It may well be that no moral motivation would be possible if, in addition to an exploratory drive, we did not also have what Panksepp (2011) calls a CARE system. This CARE system is sometimes referred to as a NURTURANCE system in Panksepp's earlier work (Panksepp 1998), because Panksepp means "care" largely in terms of care for mates and family members, especially care for the young. Later we will explore the way oxytocin and other neurotransmitters of this system trigger loving and "caring" tendencies. The CARE system also can combine with a correlative SEPARATION DISTRESS (or social bonding) system. The SEPARATION DISTRESS system is activated in an intensely negative way when an infant is separated from the mother, and in adults by grief over loss of a loved one or by complete social isolation. Both of these "social" emotion systems make mammals generally (including humans) care about the well-being of certain *specific* individuals in their environment. In male mammals, prolactin and vasopressin, as well as oxytocin (more prominent in the female), drive an intrinsic valuing of the well-being of mates and offspring. Paradoxically, the same neurotransmitters in the male lead to aggressive behavior, especially in the face of perceived threats to mates and offspring (see Chapter 2).

But there is a major problem with placing too much emphasis on these "natural" empathic emotions. Ethics is about how we treat those we *don't* know as well as those we do. The feminist philosopher Carol Gilligan (1982) therefore points out that the empathic emotion systems by themselves are not sufficient to ground the kind of universalized conceptualizations needed for mature moral conflict resolution and fair-minded social thinking. In Gilligan's view, and also that of Drucilla Cornell (1995), moral development in children requires that they eventually "universalize" their empathic sentiments. The disagreement of Gilligan and Cornell against Lawrence Kohlberg's overemphasis on abstract and "disinterested" thinking in moral development has nothing to do with any rejection of Kohlberg's demand for *universalized principles* (Kohlberg 1973, 1981, 1984). The problem is that Kohlberg essentially wants to remove affective feelings from the moral thought process altogether. But neither does feminist ethics want to remove *reason* (perhaps contrary to many popularized accounts of feminism).[3]

The crucial point is that we can't count on the naturally empathic emotions alone to motivate the universalization of moral principles. If that were all there is to it, then "You ought to do X because you are naturally motivated to do X" would reduce to "Do moral behaviors whenever you feel empathic." That is, "Do moral behaviors whenever they happen to coincide with whatever you feel like doing," or more simply, "You ought to do whatever you want."

This brings us back to the "why be moral?" problem. Why should we care about the well-being of those we *don't* know? To be sure, we already do care (for innate reasons) about some specific others – mates, family members, etc. But we also want to explore (also for innate reasons) in order to discover how much we ought to care about a larger number of sentient or valuing creatures than we might initially find ourselves caring so much about. And we want to know *how we should distribute* our efforts. We want to know how much responsibility we should take, for example, for the beggar on the street or for the sick children in Iraq where White Phosphorus explosives left dangerous cancer- and birth-defect–causing residues behind after the war ended. Equally important, we want to know how much moral responsibility we can reasonably expect of others, and under what circumstances.

But at the same time, we have other incentives to deceive ourselves at every step of the process. Our peers, the older generation, and many other factors are constantly working to seduce us away from the moral-truth-seeking function. "The beggar would have spent the money on liquor anyway," we are told. "To give him a handout would just foster dependence." We then succumb to our natural tendency *not* to pay empathic attention to the disturbing, hapless, or even potentially dangerous image of the beggar.

To motivate universalized principles, the existence of an innate and independent exploratory drive implies that the naturally altruistic/cooperative tendencies also can be *combined with* the curiosity and desire for exploration that are initiated by the SEEKING system. This exploratory tendency motivates what Hume (1740/1955, Book II) calls a "love of truth." In contemporary terms, Hume's love of truth (the exploratory drive) results from an independent and endogenous emotional brain system. Chapter 2 will explore some aspects of this system. While moral motivation may require a capacity for empathy, neither hedonistic nor altruistic reinforcement can serve the purpose of energizing the organism toward exploration – in this case, the exploration of universalized moral concepts. Otherwise, the moral-truth–seeking function would always be subverted in cases where it *conflicted* with those other interests.

An obvious problem with this emphasis on truth is that, in the realm of moral controversies and conflicts, establishing a rationally determinable "truth" is at best a daunting challenge, if not in fact a logical impossibility. Increasingly in the modern era, attempts to prove the basic principles of competing moral systems by means of logical argument have tended to be debunked by careful analysis (see Chapter 7) – leaving a vacuum seemingly to be filled only by the "natural moral emotions."

Moreover, it becomes increasingly dubious to think that moral beliefs could be based simply on religion or social indoctrination, because members of the *same* religious denomination and social grouping can have *diametrically opposite* social and moral worldviews. In fact, hardly any Jews or Christians still believe that someone who eats shellfish will have all of his wives stolen by Canaanites, as the Book of Deuteronomy clearly specifies. Moreover, the Book of Exodus unambiguously declares that those who work on the Sabbath, without exception, should be stoned to death; but here again, hardly any Jews or Christians still stone people for this reason.

The failure to show how moral statements can be "verified" – through either religion or logic – has remained a serious embarrassment to any viable alternative *other than* some form of emotivism. The classic version of emotivism is A. J. Ayer's (1946) idea that moral statements are "merely expressions of emotion." Granted, as Richard Joyce (2008) points out, a naturalistic worldview may not technically imply "emotivism" in Ayer's strict sense. People don't *mean* to be "merely expressing their emotions." But if naturalism is not compatible with some realistic notion of non-arbitrary truth in ethics, then we would still be left with emotivism as our remaining option. "Cultural relativism" is just another form of emotivism, combined with the additional assumption that most people in a given culture feel the same way. A complete cultural relativism eschews the idea of moral "truth," just as much as other forms of emotivism.

So even the idea that the exploratory drive naturally motivates a love of truth still doesn't immediately guarantee that there *is* some objective truth to be found in the moral realm. Nor does it guarantee that we would be motivated to guide our actions in accordance with such truths. Both of these problems pose serious challenges. An equally daunting puzzle is how to define "X has value" in a way that doesn't merely reduce to "I value X."

Nonetheless, it will become increasingly clear that some overall moral worldviews "make more sense" than others, because they contain fewer incoherencies in terms of conflict with empirical reality, logical self-contradictions, and the need to motivatedly misinterpret the meaning of some of our own emotional feelings in order to prop up an incoherent viewpoint. As Frijda (2006/2007), Salmela (2014), Gendlin (1982, 1992, 1997), and many others have noted, misinterpretation of our own affective feelings is a very common occurrence. Here again, these inner conflicts rage within the human brains of members of other cultures as well as our own. The fact that ancient cultures may seem alien or primitive doesn't imply that those people's brains were so different that their members would monolithically and mindlessly agree to unreflective moral assumptions any more than we do. Paulin Hountondji (1987) emphasizes this point with regard to the tendency to identify one monolithic "African philosophy." The truth was always hard to find, and agreement even harder.

By using the term "truth," I don't assume any particular theoretical understanding of or criteria for "truth" – for example, the traditional "correspondence" notion, or the phenomenological idea of an experiential relation of "disclosure" (Heidegger 1927). There have been a number of divergent philosophical attempts to define "truth." But for practical purposes, when we say "Iraq didn't have nuclear weapons," we believe it is *true* that Iraq had no nuclear weapons – regardless of which theory of "truth" we are using. And we believe it is *false* to say that they did have them.

The exploratory drive goes far beyond merely occupying a small perch atop a grand "hierarchy of needs," as Maslow (1962/1970) placed it. Instead, for White, Montgomery, and Frijda, as for Panksepp and Watt, the exploratory drive is equally important and necessary with all the other natural emotion systems. In Panksepp's analysis, the SEEKING system in the emotional brain – of which the exploratory tendency is one manifestation – may actually be in a sense more basic than the other drives. At the level of subcortical arousal and action-initiating functions, the other drives depend on the SEEKING system to energize the brain for all of our actions and to motivate the direction of attention. In Frijda's terms, there must already be an "action readiness" before environmental conditions can either afford or thwart the corresponding activity.

We often tend to think of the "search for truth" merely as a cognitive *skill set* in the service of other emotional drives. But emotional research is finally beginning to confirm that a *desire* to know the truth about the world is more than a skill set. It is grounded in an endogenous and independent *emotion system.*[4] More than just a steering wheel, it is part of the fuel that drives us. On Panksepp's and Frijda's analyses, Mother Nature in her wisdom engineered us so that this persistently driving emotional system is *not* dependent on the reinforcement of other motives. For that reason – and in my view only for that reason – the "love of truth" is not always automatically trumped by the confabulatory whims that our other emotions tempt us so often to concoct.

The desire to know the truth about our world has its own motivational energy and its own alluring glamor. It is present in many species of animals, with very similar emotional neurophysiology as in humans – even without the human intellect applying it to our wider range of often abstract questions about reality. Cats, for example, don't need to be reinforced for exploring their environment, as every cat lover knows. They will explore even if punished for doing so. The same appears to be true for humans. Hsee and Ruan (2016) use human subjects in several experimental settings to show that "humans possess an inherent desire, independent of consequentialist considerations, to resolve uncertainty; when facing something uncertain and feeling curious, they will act to resolve the uncertainty even if they expect negative consequences" (2016, p. 1). They conclude that curiosity functions independently of hedonistic reinforcement.

The change of thinking on this exploration issue is reflected in twenty-first-century zoos, which understand the importance of allowing some free range to animals, unlike most of their twentieth-century counterparts. Like lab and zoo animals, we all want to explore even if at the expense of a certain amount of risk and negative reinforcement; but if punishment includes almost complete deprivation of exploratory freedom, even the best-fed animals eventually become depressed or, as Pavlov described his burnt-out laboratory animals, "neurotic" (Pavlov and Gley 1929; see also Cook 1939). Even chicken farmers are now realizing that depriving the chickens of any free range whatever is inhumane.

Less empirically oriented psychologists and philosophers may suspect that delving into neuropsychology is not really necessary in order to understand moral psychology. I plan to show that, even from phenomenological and deconstructive standpoints, there is a very useful and even crucial role to be played by the new trends in emotion neuropsychology. In a sense, one might say that the same tools that were used to dig moral psychology into the deep bog of "anything goes" relativism may also be useful in helping to dig our way out.

1.2 Clinical Depression and the "I Prefer Not" Syndrome

As hinted earlier, the general arousal and action-energizing aspect of our subcortical emotional functioning helps explain why *value nihilism* can't be a real option for humans. Because the action-initiating aspects of subcortical function arouse us to *self-motivated action*, they undergird the natural feeling of "inspiration" that gets dampened in clinical depression. This point is going to become important in the internal-conflict approach to moral psychology, because the feeling of inspiration ultimately is needed in order to avoid value nihilism, and thus as a necessary (but not sufficient) condition to motivate moral behavior. A truth-oriented approach to ethics must build from the foundation of a rejection of value nihilism, as will become increasingly obvious. Panksepp et al. (2014) specifically link depression to a dampening of the SEEKING system, which is necessary for feelings of "enthusiasm" – similar to the mysterious "will to live" that continually astonishes physicians.

Panksepp et al. also acknowledge that there is a sadness component to depression. But while depression usually *includes* anhedonia as a symptom, simple anhedonia by itself is not depression (for example, see Gabbay et al. 2013). More basic than the failure to experience reward is the inability in depression to value any outcome enough to feel rewarded even if the outcome is achieved. The inability to value outcomes strongly enough to feel gratified even when they are achieved can be thought of as a kind of value nihilism. More than mere "apathy," in depression we don't merely *lack* feeling, but we actually feel emotionally disturbed that we can't value anything strongly enough to be rewarded when we achieve it. We may force ourselves to act, but we can't actually *want* to take those actions.

If our self-organizing action systems are functioning robustly, then most of us, however satisfied or dissatisfied our other needs might be, could never simply say to our entire lives, as Melville's Bartleby the Scrivener did, "I prefer not." In Melville's story, Bartleby simply decides that he "prefers not" to do anything whatever. Areas of Bartleby's brain must have been shut down that normally would play a crucial role in the brain system that Panksepp calls the SEEKING system.[5] We will explore a few of the details of this process in the next chapter.

Systematic shutdown of this action-motivating system thus leads to an extreme absence of any feeling of inspiration – in other words, clinical depression. Depression shuts down our motivation to act of our own volition, as opposed to merely reacting, by depriving us of all sense of "gusto" or "will to live." In terms of everyday subjective experience, the opposite of clinical depression is not happiness, but rather inspiration in this sense. Depression especially subdues our curiosity and desire to explore in search of knowledge about our environment and the world generally, which the new neurophysiology of the innate exploratory drive shows we desire for its own sake, not merely as a conditioned response.

Inspiration therefore should be listed as one of the "basic emotions," because it is always present in the background, except in extreme cases of depression. If "basic" means "unconditioned," then certainly the always-present subcortical fueling of inspiration should qualify. Even twentieth-century hedonistic theories included the need for a "general arousal" system, but didn't view this arousal as unconditionally motivating actions, nor as extending to a subjective feeling of inspiration or enthusiasm. It was therefore not even thought of as an emotion, let alone was its significance recognized for moral psychology. Granted, even a depressed person can have cortical arousal, in the sense that dopamine is being released; but the level of functioning of the overall SEEKING system is low.[6] So this same self-motivating action system can be viewed as not only unconditionally leading to a desire to explore the environment – an environment that for humans includes abstract as well as concrete features of the world – but it is also needed for the feeling of inspiration.

The "I prefer not" phenomenon suggests a basic distinction between two value dimensions. One is the reward we experience when we *achieve* whatever it is that we value. But underneath that dimension, and presupposed by it, is an equally basic one: We must be inspired to *feel the value of* that which we value in the first place, and with a certain degree of intensity. There is little point in achieving anything if we don't value anything very intensely to begin with. Actions whose purpose is to intensify these feelings of inspiration – the feeling of the value *of* that which we value – are different from the actions whose purpose is to *achieve* whatever it is that we value. The former include, for example, artistic activity, love relationships, moral action, friendly or helpful behaviors, and political activities – actions whose purpose is to give what Gendlin (1962/1997) calls "concretely embodied symbolization" to our sense of valuing something, independently of whether we happen to achieve it.

Depression in the clinical sense, which includes a nearly complete lack of inspiration for action, often can result initially from a deficit of SEEKING, and then subsequently may affect the hedonistic systems indirectly (Panksepp et al. 2014). And in some cases the causation may be reversed: Too much feeling of sadness over a prolonged period can decrease the amount of dopamine to such an extent that we are no longer able to feel motivated to take action, since dopamine plays such a crucial role in voluntary actions (for example, see Dunlop and Demeroff 2007). In that case, we lose the ability to value any action intensely enough to energize the action. We thus eventually come to feel extremely apathetic, as if we valued nothing.[7]

In either case, whether prolonged sadness indirectly de-motivates action by suppressing dopamine (and/or serotonin, in a more indirect way), or whether de-motivation eventually leads to sadness, depression is characterized more by a shortage of the intensification of our *feelings of the value* of our actions (and their aims) than it is by an inability to attain the hedonistic rewards that result when those values are facilitated. It becomes depression when we no longer feel inspired by the actions we want to take.[8]

The bottom line in all these cases is that if the feelings of sadness are not prolonged or deep enough to lead to a lack of "inspiration" in the sense we are using – Panksepp's "enthusiasm" – then the condition is only sadness, grief, or "dysthymia," not an extreme case of clinical depression. And in many cases the lack of inspiration is actually the cause of the feelings of sadness themselves.

So it does little good to try to help a depressed person achieve valuable outcomes. Before outcomes can be valued, an underlying inability to value per se must be overcome. This value nihilism stems primarily from a suppression of the SEEKING system. This valuing dimension also interrelates in complex ways with the endogenous exploratory tendency, leading to what Hume calls the "love of truth." The meaning and implications of these concepts need to be carefully unpacked.

1.3 "That" Must Not Determine "What"

Unlike other emotions, an endogenous and independent exploratory drive would motivate *that* we seek moral truth, but it would not predetermine *what* we take the guiding principles of action and morality to be. The resulting moral view would no longer read: "I should do XYZ because I am naturally motivated to do XYZ" (which I am arguing is morally vacuous). Instead, it would read "I should do XYZ if I believe it to be the morally right thing to do, and moreover, I am naturally motivated to *find out* what is morally right." The result is a non-vacuous outlook that doesn't reduce "I ought to do XYZ" to "I want to do XYZ." The motivation to act according to what we believe is right could lead to XYZ, or to anti-XYZ, depending on the results of our moral *thinking*.

I realize that this view runs contrary to at least a century of hedonistic psychological theory – and the increasing cultural consensus that we are all just complicated hedonists at heart. But the new emotional brain research is explicitly going beyond the old hedonistic reinforcement theories. Mounting evidence now challenges the vision of humans as the pre-consciously calculating yet socially programmed reinforcement-seekers that twentieth-century psychology tended to suggest, even granted that empathy and fellow feeling were counted as "reinforcing" in the hedonistic approach. Increasingly, physiologists realize that some of our instinctive motivations involve behaving in certain ways independently of whether doing so is reinforced by the expectation of the food pellet or shot of dopamine at the end of the maze. In recent years, emotion neuropsychologists take one of these types of behavior to involve exploration and curiosity, which I will argue also extends to the realm of ethical beliefs and overall philosophical worldviews.

There is a double irony in the fact that twentieth-century psychology almost succeeded in bringing philosophical ethical theory to its knees. The first irony is that the increasingly scientific approach in academia wanted to foster objectivity and impartiality as against the nice illusions we might prefer to believe. But scientific psychology and the other social sciences ultimately tended to imply that moral beliefs *cannot* be rational or objective. Theories of psychological conditioning constrained them to assume that the only check against brute egoistic hedonism (including the indirect hedonism of the nurturing parent) must be found in blind social conditioning. There could be no inherent natural motivation to act contrary to basic hedonistic interests, broadly construed.

By "broadly construed" and "indirect hedonism," what I mean is that the human mother's defense of her cherished infant was taken to be just as ruthlessly amoral as that of any mother lion or tiger. It was driven simply by her own organismic and ultimately hedonistic motives, and had no more epistemic validity than that. In the end, the only natural check against the hedonistic tendency was either *mindless social convention* or *innate altruistic feelings* toward specific individuals and groups – and both these forces engender the very opposite of objectivity for any non-trivial ethical belief system.

For example, Kathleen Rankin (2007) observes that sociopathy includes an inability to experience the "social reinforcements" that would be necessary to guide moral behavior for the normal person (necessary, but notably not sufficient). As for the philosophical interpretation of what is going on in normal, *non*-sociopathic moral behavior, we should not leap to the assumption that ethical behavior results only from such social reinforcements. We can't infer from the sociopathic deficiency that moral beliefs and behavior are determined purely by the organism's need to maximize hedonistic rewards (both physical and "social" rewards). If they were driven only by these "rewards," then ethical beliefs and behaviors would become a secondary byproduct of this hedonistic system. In short, we would believe whatever gives us the most pleasure to believe. In terms of the grounding for moral thought processes, we would then be very far from the eschewal of irrationality that scientific thinkers might initially have hoped for.

This first irony of twentieth-century moral psychology led to a second one: Value systems not only became even more arbitrary and less rationally based than before; in a sense, many otherwise intelligent people quit taking moral reasoning very seriously at all. Sheer hedonism could hardly support feelings of inspiration – the feeling that writing a symphony or acting by a moral principle is worthwhile for its own sake as opposed to merely an indirect means to maximizing pleasure. So the more life was geared toward gratifying one's own natural drives, the more quaint and absurd the problem of the "meaning of life" came to sound. Consistently with the limits of empiricist epistemology, ethics became more and more a "descriptive" rather than "prescriptive" enterprise. All values and value systems were equally lacking in objectivity, and could not *really* be believed, regardless of how much they were socially conditioned. Everyday people on TV talk shows routinely assumed that value beliefs were "just a matter of opinion" or "how someone feels about it."

But if we really believed that all non-egoistic values were mere social conventions, we could hardly feel inspired even by the conventions of *our own* society. To the extent that we believed in science, we had no reason to be inspired by any particular value system. This notion that ethical thinking at bottom was arbitrary then made us even more susceptible to the moral self-deceptions motivated by egoistic hedonism – the tendency to believe whatever is most consistent with our own self-serving ends – again, just the antithesis of scientific objectivity. Deprived of the inspiration offered by the valuational realm, many people began to feel that their lives, however pleasant they might be, "lacked meaning." Yet we couldn't frame this question of "meaning" in the terms that scientific psychology offered us.

It is crucial in this context to distinguish between moral *relativism* and moral *relationalism.* Obviously, moral judgments are "relational" in the sense that the same action may be right or wrong in different social circumstances. Moving my fist may be wrong if someone's face is in the way, but not wrong in other circumstances. Similarly, picking crops on someone's land may be harmful if done in a context where economic and social exchanges depend on rules of property ownership, but not harmful in a society where property ownership is not part of the agreed-upon economic system.

But as Gestalt psychologists like Karl Duncker (1939) point out, this issue is unrelated to the question of ethical *relativism* – whether it is actually true or merely an emotional feeling or arbitrary social convention that X is wrong under such-and-such circumstances. For example, we can still ask whether moving my fist *really is* wrong in a situation where someone's face is in the way, or that picking crops on someone's land sometimes *really* may be wrong if the owner's economic survival depends on use of the property. As Edward Ragsdale (2006) points out, the *meaning* of an action – for example, a woman removing her veil – may also be different depending on the cultural context. So it is not necessarily true that the same action *with the same meaning* could be right in one society and wrong in another. The meaning "nepotism" in one culture might be similar to the meaning "networking" in another culture. "Bribery" in one culture might mean approximately the same as "lobbying" in another – especially when lobbyists can spend unlimited amounts to defeat any legislator who "disagrees" with them. Similarly, a European or American woman showing her ankles in the nineteenth century has a different meaning from the same behavior in the twenty-first century.

Granted, it follows even from "relationalism" that to suddenly change a long-standing system of social customs without careful consideration can sometimes cause some degree of harm, even if "harm" is defined *in objective terms.* In the Middle East, for example, there is an interconnected system of customs associated with extended-family living – the father ceremoniously brewing the espresso coffee for the entire extended family, the male cousins accompanying their female relative in public places, the openly nepotistic hiring practices, etc. Tearing apart that interconnected social fabric could have unintended harmful consequences when we notice that the rates of murder and suicide within that long-standing cultural tradition are much preferable to the astronomically higher rates in the Western lifestyle – especially in the United States, where murder rates are 10 times as high and suicides rates 20 times as high (Eshun and Gurung 2009). The need to weigh the harm of social upheaval and *anomie* against the harm caused by the system itself doesn't imply *relativism,* although it does imply *relationalism.*

No philosophical problem results from the social sciences stressing relationalism; the problem occurs when we confuse relationalism with relativism. But a genuine relativism (not just relationalism) was one of the unintended consequences of the over-emphasis on *hedonism* in psychology, since it seemed to reduce ethical belief to mutual egoism, natural altruistic feelings, or some form of mindless, socially conditioned reflex.

Against this backdrop, enter the new emotional neuropsychology of the twenty-first century, in which the exploratory drive is neither clearly hedonistic nor completely egoistic. Nor are the non-hedonistic impulses exhausted by either the "empathic" or the "culturally constructed" emotion systems. We should acknowledge that the roots of the new psychology of emotion were already present in neglected aspects of twentieth-century research and theory, tracing as far back as Sigmund Freud's *Beyond the Pleasure Principle* (1925) and Otto Rank's *The Trauma of Birth* (1936). But only recently have non-consummatory (i.e., non-hedonistic) drive theories become prominent and empirically well-established. In light of the newly appreciated emotional neuropsychology, the crucial implication is that *ethical beliefs* need not be pre-ordained by any combination of emotional preferences – not even hardwired *altruistic* preferences. Less still need they be reduced to behavior patterns conditioned by social reinforcement.

To be sure, the question as to how an emotional act of "valuing" connects with a non-subjective statement about what kinds of things *"have* value" will become important in the approach we are taking. The problems involved in defining what people *mean* when they make value statements that are not intended as mere expressions of emotion will have to be dealt with. We will reserve a special place for those problems in Chapter 4 – after some of the crucial issues of inner conflict have been broached.

We can also tie the new trend in emotion neuropsychology to phenomenological approaches such as those of Maurice Merleau-Ponty (1945/1962, 1942/1963), Martin Heidegger (1927/1962), Max Scheler (1954/1970), Rollo May (1969), and Eugene Gendlin (1997, 1978/1982). When phenomenologists reflect into the subjective side of experience, what they find is not merely hedonism. Scheler, for example, observed that we value the well-being of others "intrinsically" – i.e., for its own sake – not just as a means to our own ends. We can't feel that a friend's welfare will no longer matter the day after we die, as if the friend's value stood in need of our own value feelings to ground it. Levinas (1969) reaches a similar conclusion.

In still another application of phenomenology, Heidegger as well as his diverse students Hannah Arendt (1959, 1963), H. G. Gadamer (1971/1975), and Hans Jonas (1966/2001) focused on the notion that our caring about *future action potentialities*, in a temporal structure that can motivate action, affects the *direction of our attention*, and thus eventually our *interpretation of reality.* For Heidegger, our assessment of the "actual" world, including questions about value and the meaning of life, are thus dependent on this action system motivating us to care about future potentialities.[9] This approach is consistent with the notion that action and not mere reaction – i.e., self-organization – is key to the value dimension as applied to moral thinking and motivation.

We will also find that phenomenology can help with the connection between "value" and "meaning," which is more complicated than a simple identity between the two. Experiencing our actions and lives as "meaningful" is not just a matter of attaining something that we value, or even valuing the right things. Meaning has to do with the intensification of the *experience of* value more than with its achievement. Real actions in the world have to be motivated or inspired in some way, and this inspiration dimension must be attended to in order for actions to have "meaning" in this sense. This will also imply a continuity between "moral" and "nonmoral" values, as will become increasingly obvious.

I am aware of the usual phenomenological skepticism regarding the "naïve realism" that tends to be associated with empiricist investigations. But I will argue that, in the spirit of Merleau-Ponty and Jonas, we need to integrate empirical *with* experiential understandings of embodied consciousness, since it is through the lived body that we experience the world. The brain is part of the body. In a sense, it is even part of the *lived* body in that we use brain processes to subserve the activities that we "live." In the view that I want to argue, the brain/body system encounters and understands the world by acting upon it rather than just reacting to it.

1.4 Why the Naturalistic Fallacy Is Still a Fallacy

The independence of the exploratory tendency implies that moral action (like other kinds of action) is not just a *re*-action to external stimuli. It is the internally energized action-initiation of a self-organizing dynamical system (as in the new wave of "enactive" cognitive theories exemplified by Newton 1996, Varela et al. 1991, and others to be discussed later). This self-organizational basis for distinguishing action from mere reaction will become increasingly important as we think about what motivates not only moral truth-seeking behavior, but also the tendency for the resulting moral beliefs to motivate action.

When we refer to an exploratory drive, we are not referring to the prefrontal brain areas that are needed for the *skill* of truth-seeking, but rather the *motivation* to seek the truth, which is more subcortical and limbic. Because this emotion system is relatively separate and innate, it directs the normal human brain, as our cognitive functions develop, to search for the truth about the world somewhat independently of how we feel about things or what we would *like* for the truth to be. Those other desires and feelings are orchestrated by their own different brain systems.

What makes things interesting is that this desire to know the truth, as endogenously motivated by the exploratory drive, is continually in conflict with other basic motives that lead to a tendency to believe things congruent with what we would prefer to believe – regardless of evidence or logic. So the point is not to choose which of these opposing tendencies represents "essential human nature." We shouldn't say "Moral and political beliefs are basically rational," or "Moral and political beliefs are basically irrational." Instead, we need to explore the playing out of the *conflict between* equally important kinds of motivational systems.

It would be a mistake to regard this conflict as a playing off of reason against emotion. That would render "reason" motivationally impotent and thus inevitably a mere slave in the service of "the emotions." On the contrary, the clash is a running battle between *different driving motivations* with their own special emotional brain systems and energy mechanisms.

The internal conflict between these different and independent systems becomes increasingly severe in an age when statements are so often uttered, on TV or in newspapers, or in our business and personal lives, depending on the social and economic advantages to be gained by the statement, without regard to truth value. We are told that Iraq has nuclear weapons or not, depending on the interests of the players involved; the media choose to emphasize relevant information or bury it on page 32-C, depending on their own needs for entertainment value, access to the politicians, or simply to cater to the preconceived opinions of their readers or viewers. The "experts" put further pressure on the media to "spin" the truth depending on the political interests of those "experts." On the internet, we are even bombarded with fake "news services" deliberately disseminating objectively false information. The temptation to shop around for "alternative facts" becomes ubiquitous. What kinds of internal turmoil might result when a naturally truth-seeking animal must adapt to such a way of life?

Other societies and historical eras obviously exhibit their own truth-suppressing tendencies. Imagine living during the time when the Pope could burn Giordano Bruno at the stake over disagreement with the Church's geocentric scientific views. A huge conflict at the *psychological* level – not just the social level – is reflected by the conflicting motives of truth-seeking and other needs that give rise to both medieval religious dogmatism and modern TV and "social media" culture. These same conflicts are universal throughout all periods of human history and prehistory.

The central point of John F. Kennedy's *Profiles in Courage* is the difficulty of this inner turmoil. Each of his five protagonists had to pay a small or large *price* for doing the right thing despite popular and long-engrained prejudices and preconceptions; they could easily have backed down by tricking themselves into a false view of their situation and responsibilities. Their resulting actions were not necessarily *hedonically rewarding*, but they were *inspiring* (an equally important but separate motivation), and also were congruent with those protagonists' "love of truth."

So while a drive to exploration (and thus truth-seeking) does exist, it is always potentially in conflict with many other emotional and survival interests. Correlatively, it is a truism of daily politics that we are susceptible to being deliberately misled through skillful exploitation of various emotions that naturally conflict with the truth-seeking drive.

Given the effects of an independent exploratory drive, the ethical truth-seeking function doesn't need to be explained through reinforcement based on some hypothesized indirect connection to the general "benefits of social cooperation" – which presumably would represent a mutually egoistic arrangement. The Nazis of the Third Reich developed an excellent system of "social cooperation." And as Hannah Arendt (1963) was so widely scorned for pointing out, the administrators of Nazi concentration camps could be quite pleasant fellows when they were motivated by their "altruistic sentiments." So too were many of the usually charitable citizens who established the brutal sixteenth-century Jewish ghettos in Rome, Madrid, and many other European cities. Even the current perpetrators of war crimes around the world, such as genocide and the group rape of defenseless women, believe they are acting out of altruistic concern for the benefit of certain groups (while of course harming others). And members of ruthless crime syndicates often are nice people when they *want* to be.

Despite these facts, many sociobiologists (for example, Dawkins 1976/2006) try to ground ethical behavior purely in naturally altruistic and/or cooperative sentiments. The usual result is the injunction "You should act altruistically because you are naturally motivated to do so." A slight variation of this view is that "You should act altruistically, and by the way, you *also* are naturally motivated to do so." But philosophers had already realized, in a tradition tracing back at least as far as G. E. Moore (1900), and arguably even to Plato's *Republic*, that this statement is essentially equivalent to "Act altruistically when and if you want to," which then reduces to "Always do whatever you want." Even if something in the human gene pool drives us in certain instances to be cooperative or nurturant toward other humans (as is persuasively argued, for example, by Bowles and Gintis 2013; Filonowicz 2008; Roberts 2013; Ben Ze'ev 2000, 2002; Sober and Wilson 1998; and the various authors in Sussman and Cloninger 2011, as well as Sinnott-Armstrong 2008), the result would still be that we would behave in those ways *when and if we want to* (i.e., are naturally motivated to do so). But the injunction that we should do whatever we want, whenever we want, has no non-trivial ethical content and especially fails to provide guidance at times of moral soul-searching.

The term "naturalistic fallacy" was initially coined by G. E. Moore to describe this type of problem. In effect, to assert that we should act ethically because we naturally want to is the paradigmatic example of Moore's naturalistic fallacy. To describe the way humans naturally behave provides no motivation to behave better than we naturally were already going to behave. To tell someone that they always ought to do *whatever they already were going to do anyway* provides no help in making moral decisions. If religious animosity is as "natural" as ecumenical harmony, then how can a naturalistic approach ground an injunction that the latter is morally better than the former? Naturalism, obviously, must avoid moral injunctions *lacking* in natural motivation. There can be no actions without motivational processes to produce them.

It does no good to object that both moral and immoral actions are "equally" naturally motivated. The problem is that, as soon as we adopt a naturalistic perspective, we are committed to the notion that we are going to do whichever behavior is *more strongly* motivated. And the resulting danger is the implication that we will do whatever we most want to do, whether moral or immoral.

Ultimately, I will argue here that what we naturally "want" is not merely to engage in this or that particular altruistic behavior toward some specific individual or group. We also want to *find out* what the truth is regarding a basic set of moral *criteria* for what we should do, and then act accordingly. This approach is consistent with the central point advocated by Jesse Prinz (2011). He emphasizes that "emotivism" should not be confused with what he calls "emotionism." "Emotionism" is the belief that there could be no moral values without emotions to ground them; but this is not the same as saying that moral values are simply reducible to emotional preferences. So Prinz believes that "ethical realism," the idea that there are objective moral truths, is compatible with the fact that all of our moral behavior must be motivated by natural emotions. My project here is to show how this is possible without allowing natural emotions to dictate which moral principles we end up following. The key to this possibility is the independence of the exploratory drive. And the explanation for why we so often fail in this regard is that there are conflicting basic emotions, some of which motivate us to suppress our desire to know the truth in certain specific circumstances.

These internal motivational conflicts don't imply an automatic rejection of either naturalism *or* the actual truth and falsity of ethical claims ("ethical realism"). Here again, the independent exploratory drive offers a new glimmer of hope for a compatibility between naturalism and ethical realism. I refer to "ethical realism" as the view that moral statements can be true or false independently of how we feel about them – that they are not merely expressions of emotional preference. Ethical realism, for example, would hold that the other person's well-being would have had value *even if I had never existed* in order to "feel" that it had value. I plan to argue that an ethical realist perspective not only is compatible with naturalism, but also is more coherent than non-realist perspectives such as strict emotivism – the idea that ethical statements are merely expressions of emotional preference or social convention. But the naturalistic fallacy dilemma must be faced, whether one takes a realist or non-realist position.

1.5 Veneer Theory versus the Neo-Amygdala

Some may wonder why we can't simply posit that social conditioning toward altruistic or cooperative behavior has survival value *for the species* (Dawkins 1976/2006; Singer 1981; Wilson 1975/2000). After all, isn't it often in the survival interest of the group to teach its individuals to sometimes act contrary to their own *individual* interests and desires? The idea that social conditioning is the mechanism for this effect is sometimes referred to as "veneer theory" (Frans De Waal 2006, 2008). On this view, morality is a "veneer" covering over our more basic brute-hedonistic nature. Moral principles could be a general part of such social indoctrination.

But in the first place, this move presupposes an almost infinite ability of "nurture" to override "nature." If morality had to rely on social conditioning to consistently suppress our genetic nature, it is difficult to imagine why only about one to four percent of the population end up as true sociopaths (one percent in Eastern Asia as opposed to four percent in the United States, according to Stout 2005). Only those tiny percentages actually do behave as pure egoistic hedonists. How is it possible that social remolding of a hardwired "brute nature" could succeed almost universally in suppressing an innate egoistic hedonism, even though the *motivation* for the social veneer is supposedly *hedonistic* in its own right? If morality were a result of social indoctrination, then wouldn't we have more sociopaths than we do, in all the cases where the social conditioning had been ineffective? If morality depended on social conditioning, we should expect that many would remain in the supposedly "natural" original state of sociopathy. Correlatively, if morality were socially conditioned, it should be easier than it actually is to socially condition a sociopath to *become* a morally functioning person.[10]

Even those few who do end up as sociopaths don't fail to develop a moral sense primarily because of poor child-rearing practices or lack of trying by the social community. Sociopathy is notoriously resistant to good parenting. In fact, recent research suggests that the "acquired" sociopathies may very well result primarily from a physiological brain condition caused by orbitofrontal brain injury, which leads to worse outcomes the earlier it occurs during childhood; whereas "true" sociopathy (that is, not "acquired") can result from a developmental abnormality of the orbitofrontal-amygdala system during infancy or fetal development (for example, see Anderson et al. 2009; Faw 2000a, 2000b, 2003; Kim et al. 2007; Koenigs and Tranel 2006; Rankin 2007). Of course, some of these brain abnormalities could be caused simply by severe physical *beating* of the head by abusive parents.

Studies show only very modest correlations regarding sociopathy between siblings raised together, and only weak correlations between twins raised separately (Hicks et al. 2004). But in most twin studies, the twins were not separated at birth or even during the first few years, whereas the brain injuries most likely to cause the most serious acquired sociopathies (as well as severe impulse control issues) occur in the first three years (Anderson et al. 2009; Faw 2015; Kim et al. 2007). So even the very weak correlations that do exist with twins raised separately don't necessarily prove that the cause is always genetic rather than a result of early brain injury. Here again, the injury sometimes could be caused by severe beating, and sometimes by a mere accident. What is clear is that neither sociopathy nor "acquired" or "pseudo"-sociopathy is primarily or usually caused by the parents' social conditioning (Hicks et al. are very explicit on this point). All things considered, evidence suggests very little support for the idea that people would "naturally" be sociopaths if it weren't for a social "veneer" overlaying a more basic brute egoistic hedonism.

By contrast, there is plenty of evidence suggesting patterns of *natural brain growth* that subserve normal moral development. Recent brain studies (for example, Gee et al. 2013; O'Doherty and Dolan 2006) show that, during later childhood, new brain mass develops *around* the "old amygdala" (which we share with other mammals). Bill Faw (2015) refers to this new area of amygdala as the "neo-amygdala." The "neo-amygdala" is a large brain mass that literally surrounds the "old amygdala," and it connects to the orbitofrontal cortex, which is involved in critical thinking and impulse-regulation (Chareyron et al. 2012). The old amygdala in early childhood triggers an immediate, unreflective emotional response (Tottenham and Casey 2009). But in later childhood, the old amygdala is not the first emotional area to receive activation from a salient stimulus. Instead, the *neo*-amygdala-orbitofrontal circuit is activated first (Barbas and Zikopoulos 2006; Faw 2015). If the adult's or older child's decision then is to react with intense emotion (for example, to fight or run away), *only then* is the old amygdala hyperactivated in order to energize the action more strongly (O'Doherty and Dolan 2006). In other words, the neo-amygdala acts almost as a *shield* for the old amygdala, preventing it from hyperactivating at the drop of a hat. In this sense, it is more "natural" for the adult or teenager *not* to fly off the handle than to do so.

So the idea that we must in real time "suppress" what our old amygdala naturally does in order to regulate our behavior in a socially acceptable or prudent way is an outdated notion. The development of the neo-amygdala is a purely physiological phenomenon resulting from a natural biological pattern of brain growth, and is not primarily a result of social conditioning – although of course its effectiveness can be tweaked by learning. (As discussed in Tottenham and Casey 2009, the ways we habitually *use* our brains do somewhat influence their physical organization – a point also stressed in Ellis and Newton 2010; later we will discuss the crucial importance of learning to regulate the fear responses of the old amygdala, which tend to inhibit rationality.)

The "wild boy of Aveyron," contrary to popular opinion, probably was not raised by wolves, but he did spend at least some of his childhood isolated from other humans. Even though he was unable to acquire human language, he did exhibit a fairly normal moral concern for people with whom he interacted (Aroles 2007; Bettelheim 1959). He didn't need to be *taught* to be moral, as though this would be a social remolding of an otherwise amoral natural condition.

If sociopathy is not the genetically natural condition for the normally developing brain, then morality is not purely a result of social conditioning. Correlatively, morally relevant normal brain development involves more than just the so-called "CARE" (or nurturance) system contained in the mammalian brain, and it requires more than an innate social bonding system (Panksepp's "SEPARATION DISTRESS" system, discussed in

Chapter 2). Both of those systems also have to interact *with* the SEEKING system in the emotional brain in order for mature ethical thinking, deliberation, and inspiration to take place. (Parenthetically, empathic emotions also require a normal "mirror neuron system," as discussed later.) The *interaction* of these different emotional brain systems naturally leads to moral development in very intelligent mammals (for example, humans), given minimal social resources. But here again, this natural moral development is often in *conflict* with other equally natural motives that lead us to tweak our interpretation of reality to justify non-moral or even immoral behaviors in the service of other needs and desires.

This internal conflict is why, despite our moral tendencies, humans famously can be and frequently are among the most ruthless and destructive of all animals, as discussed in such horrifying detail by Charles Harvey (2017). The violence and devastation wreaked by humans are probably the main reason sociobiologists have tended toward "veneer" theory. But the neurophysiological considerations just mentioned are one counter-indication against such a "veneer" theory. Harvey notably does not endorse veneer theory, partly because he points out that modern "civilized" humans have now become if anything even more destructive than our prehistoric counterparts. As a simple example, the destructiveness of human warfare over the long course of our history and prehistory has dramatically increased, given ever-more-sophisticated technologies, and also the paleological evidence about prehistoric warfare and the lack thereof (for example, Guthrie 2005; Kelly 2000; Olson 2002). Those who depict early humans as equally warlike with moderns (for example, Keeley 1997) are not looking as far back as 20,000 years ago, when the trends toward warfare, overshooting of population niches, etc., had already become more pronounced.

So Harvey's point is that the "veneer" of civilization has by no means made the human species less destructive. And if there is sometimes moral progress – as does seem to occur in some ways, such as in the recent general rejection of Nazism, racism, sexism, and homophobia – the progress occurs at least as much *despite* as *because of* the previously prevailing "social veneer."

In fact, still another objection to the notion that moral principles are entirely socially learned is the puzzle of how to explain cultural change – as for example when US culture decided to reject the racist value system (to put it bluntly) into which the vast majority had been socially indoctrinated up until the 1960s. To be sure, desegregation of schools eventually made it easier for young people to appreciate the humanity of children of other races and nationalities. The mores conveyed by Hollywood movies made it easier still. But social developments like the desegregation of schools (to the slight extent that this was de facto achieved) and the changed messaging of films had already required moral courage by earlier generations of abolitionists and civil rights workers. So the question remains: What drives each new generation to believe that the truth is different from what people had been reinforced for believing?

Some might answer this question by noting, as Otto Rank (1936/1978) and other psychologists have emphasized, that there is a natural need to define ourselves as sharply distinguished from our parents. We see this tendency in the two-year-old "no" stage, as well as the "adolescent rebellion" phase. But this dynamic by itself still would not explain why such "rebellions" so often seek (with whatever degree of success) to find more veridical ethical viewpoints rather than simply other equally wrong-headed or arbitrary ones – *unless*, I will argue, the new generation is selectively suppressing its exploratory drive.

As a result of the independence of the exploratory drive, the answers eventually reached regarding moral questions are not merely a conditioned response based on social reinforcement. Different people will reach different conclusions because their reasoning is different, even if all are driven (partially) by a motivation to seek the truth.

I am not suggesting that each generation completely ignores the accumulated wisdom of previous generations in order to construct a completely new belief system – any more than we ignore the accumulation of geometric thinking when we want to understand why the Pythagorean theorem is true. The point is that we don't accept the Pythagorean theorem simply because we are told it is true. We accept it because it makes sense to us. If we exert the effort, we can even understand the logic of the proof. The accumulated work of earlier generations of mathematicians saves us the work of having to completely reinvent the wheel. We no longer have to *discover* the Pythagorean theorem, but we do have to understand why we should believe it is true.

Unfortunately, ethics is not a completely a priori discipline like math, and therefore it can't enjoy the certainty of mathematical theorems. I will argue later that it is more like natural science in this regard, and therefore involves comparison of the epistemological coherence of a variety of different viewpoints.

1.6 Inner Conflict as a Source of Epistemic Incoherence

Moral behavior, as Stout (2005) points out, often may actually have *negative* hedonic value – as compared with the sociopath, who suffers no pangs of conscience. But the consolation for this unfortunate aspect of ethics is that moral values also have the equally important capacity of *inspiring us to action*, by contrast to Bartleby's "I prefer not" response. For an inherently active creature, inspiration – the opposite of clinical depression – is crucial. Other animals can be inspired simply by the beauty of nature, a chance to play, or the call of new territory to explore. For humans, the new territory needs to include abstract as well as concrete phenomena. The pull of an opportunity for effective socially beneficial action (I will argue) can be among the most inspiring of all action potentialities, regardless of hedonic consequences. The resulting inspiration can be seen in the patterns of activation in midbrain, orbitofrontal, and anterior cingulate areas of the brain that participate in the SEEKING system (Panksepp 1998, ch. 6; Panksepp et al. 2014).

For this reason, the natural need to think about moral issues is intertwined with the question, for humans, of "the meaning of life." In each instant we form a feeling as to what we "ought" to be doing at that moment, both "morally" and "non-morally," with regard to long- and short-term values. Moreover, these values automatically include a negation of the "I prefer not" attitude, thus an everyday feeling of inspiration that supplies the motivation to live and to act. What are often called (with deceptive simplicity) "moral" and "non-moral" values must be capable of being weighed against each other in importance in order to choose a course of action.

As we think even in normal adult laymen's terms about how to integrate the facts about our valuational experience into an overall moral worldview, we find ourselves trying to maintain epistemological coherence. The simplest *but by no means the only* example of this process is that we try to avoid beliefs that entail logical self-contradictions. The racism example, again, provides a simple illustration of this point. The predominantly white culture of the United States for several centuries attempted to believe in democracy and universal rights while at the same time relegating Blacks – and to a lesser extent Jews and recent immigrants – to unequal status and disenfranchisement. When people were motivated to change their belief system as it affected this issue, they were not just reacting to moral indoctrination from the previously existing culture. They were acting to create change as a result of moral thinking.

To be sure, the new generation was carrying forward some principles from the past and trying to make them more coherent, as Alisdair McIntyre (1984) emphasizes. But this doesn't change the fact that something happened to reduce the normal *resistance* previously associated with asking a certain subset of reasonable questions about an entrenched worldview. Part of this thinking involved the realization that the previous worldview was logically incoherent – a realization that was helped along by means of televised civil rights demonstrations continually directing *attention* to the logical contradictions of segregation-within-democracy. When people pay attention to their own logical incoherence, there is a tendency to recoil from it. This role of attention, especially the voluntary control of attention, will become increasingly important here.

I plan to show that avoiding self-contradiction is *only the most elementary requirement* for a coherent worldview. As such, the principle of non-contradiction is only the tip of an epistemological and psychological iceberg.

1.7 The Striving for Coherence – More Difficult than it Sounds

The idea of a "coherence epistemology" in ethics was broached in a primitive way by the suggestion of Norman Daniels (1979) and Michael DePaul (1986) that what is needed for epistemological coherence is a "wide reflective equilibrium," in which we try to balance all our value beliefs against each other without self-contradiction. Such a process, they believed, could lead to an overall viewpoint that is more likely to be true than a less coherent one would be. A belief in democracy is more likely to be true if not combined with a belief in unlimited power of well-funded lobbyists over legislators, since the two beliefs contradict each other.

But as Geoffrey Sayre-McCord (1985) and Gilbert Harmon (1986) point out, two difficult problems must be solved in order for such a "coherence" process to work:

(1) The criteria for coherence would need to be expanded much further than merely logical self-consistency. There simply are too many extremely divergent and controversial *yet logically self-consistent* moral worldviews. And, as in the slavery example, it is often possible to eliminate even a logical self-contradiction if one is willing to embrace just the right ad hoc hypothesis (for example, that the slaves were not "fully human"). Coherence therefore requires more than a mere absence of self-contradiction.

(2) We also must guard against the danger of even *beginning* any moral inquiry with "moral intuitions" (even if those are later to be refined with tests for coherence). Moral intuitions are too inextricably enmeshed with the prejudices of our childhoods. For this reason, mature moral reasoning can't merely begin with "intuitions" and then try to refine them so as to make them logically consistent with each other and with empirical reality. The result may be mere casuistry.

What seems clear, however, is that *some sort of coherence process* must be at work, because we can't logically prove the truth of an ethical proposition – especially one that has practical importance in the realm of moral conflicts. If we could logically prove them, they would become tautologies and thus lack real content; they would simply restate the arbitrary definitions of the moral terms that we initially set up – for example, "good" = "pleasure," or "right" = "conducive to maximizing the general happiness." Such arbitrary definitions merely stack the deck in favor of our preferred way of thinking and thus become question-begging in the most serious sense. We will have to confront these epistemological problems in later chapters. Ultimately, we will have to spell out a coherence approach that does more than just render "common intuitions" logically self-consistent.

Even if there is a natural truth-seeking motive (in conflict with other motives, of course), two major questions remain. First, how do we connect the concept of "value," which seems derivative from an emotional act of "valuing" something, to the idea of declarative statements that can be true or false – for example, "It is better (i.e., more valuable) to let a guilty man go free than to convict an innocent man"? How does a declarative as opposed to simply emotive concept of "value" come about? "Valuing" seems to be primarily an emotional act. This question can't be answered until after we lay further groundwork in the next two chapters.

Secondly, how would we be motivated to *act* in accordance with ethical beliefs once we have arrived at them? Why not simply acknowledge that there are ethical principles, but then shrug them off with a casual "so what?" – as in the so-called "acquired sociopathies" that result from adult brain damage? The SEEKING system also offers resources to help with these problems, as preliminarily outlined in the next section.

1.8 Self-Organization and Action

Action requires that we value something. This deceptively simple fact leads to extensive complications in our attempts to form coherent moral and valuational worldviews. Even if we are clinically depressed, we can't take the position that we don't value anything, try though we may.

Higher animals are inevitably driven to *self-motivated action*, not just reaction. This is an important implication of the independence of the SEEKING system, which endogenously energizes exploratory behavior as well as the arousal, attention, and self-movement processes. When an object O facilitates my desire to do action A, O is not necessarily the primary "cause" of A; as Frijda notes, the motivation to do A often is *already* there due to the organism's self-organizational structure, and object O may be only a means to achieving A. One puppy's play posture doesn't *cause* the other puppy's desire to play, even if seeing it does *trigger* an already-existing desire to play – in Frijda's terms, an "action readiness."

Self-motivated action is not a privilege of prosperous cultures, or a rare motivation at the top of a "hierarchy of needs" – a luxury to be indulged when and if we have the leisure and resources. Self-organization is *the* central feature of living organisms; it is literally the main thing that defines us as animate rather than inanimate. As a result, poor people as well as rich ones formulate hopes and dreams and then live in such ways as to try to achieve them. Serious moral thinking occurs even in the most economically depressed areas of the world. It is a common observation that the poorest people actually tend to be the most charitable. Music is played not just in "leisure" cultures, but also in places where nobody can afford to pay the musicians.

As soon as we value anything (as we must), we find ourselves trying to work out a coherent scheme of valuations that naturally includes at least some others' interests as well as our own. Even before any moral reasoning begins, we already value at least some people's well-being *intrinsically* and not just *instrumentally;* we don't feel or believe that their value is *contingent* on our act of valuing them. We can't honestly feel that a friend's well-being had no value before we met them, simply because we hadn't yet met them. We then naturally wonder about the criteria for *whose* interests should receive *how much* of our valuational energy and attention. This question motivates the intellectual work of trying to work out a coherent moral worldview.

But why should we take value statements in a "realist" sense rather than simply as expressions of emotion? The main reason stems from the notion that the way we know that a reality is independent of our arbitrary whims is that we first try to engage in actions, and then we see how independent realities resist our attempts. A realist approach to moral thinking is therefore suggested by the fact that the coherence of our moral belief systems is not subject to our arbitrary whims, but instead resists our attempts to shape it as we might prefer. This resistance implies a reality independent of ourselves. As enactivists like Newton (1996) argue, we understand both physical and interpersonal reality by noticing (and imagining) which actions we are able or unable to execute in the real world – including the worlds of math and logic as well as the physical world. At the risk of simplification, we can paraphrase Scheler (1954/1970) by saying that reality tends to "resist" our actions. In the same way, the attempt to formulate a *coherent worldview* "resists" us.

Paul Jennings only slightly exaggerates this point in his amusing 1948 spoof of existentialism, "Report on Resistentialism." Jennings defines his humorous philosophy of "resistentialism" with the pithy aphorism, "Things are against us." The president of the Society for Resistentialism opens a session by declaring "Gentlemen, Things are terrible." But there is a serious idea behind Jennings's parody. Reality is what prevents us from doing whatever we want whenever we want. On the enactivist view, exploratory behavior itself is a form of self-energized and self-organizing action, followed by observing whether we are able to carry out the attempted actions. This remains the case even if the actions involved are only *imagined* actions and thus involve physical movement only on the part of the action-imaging functions of our brains (see Ellis 2005; Ellis and Newton 2010).

We learn what is real and external because external reality resists the actions we can't perform. A real rock stops the movement of our foot. A real mathematical fact stops us from thinking that $2+2 = 5$. Even when we merely *imagine* moving through a rock, we still expect that in reality the rock would resist us, and this resistance defines its "solidity" for us. In science, we even *define* things in terms of operations we can and cannot perform – "operational definitions."

In a moral context, when we attempt to believe that all enemy civilians are actually combatants in disguise – which would make killing them easier – the actual facts of the situation resist us. Obviously, we often make mistakes in our assessment of what gets in the way of our actions. Sometimes it is only our own fears and anxieties that impede us. But to the extent that we attempt to understand external reality, we are *attempting* to distinguish it by the fact that it either resists or facilitates ("affords") possible actions.

Just as a rock blocking our movement provides objective feedback, so our attempts to adopt a coherent *moral worldview* may provide guidance precisely by resisting the attempt. When the world resists, we know that the situation is not completely a result of arbitrary subjective edicts. The same is true when the project of constructing a coherent belief system resists our attempts. We can't just believe what we want, even in the realm of ethics. When things "resist us," they are not merely the results of our emotional preferences.

When we attempt to believe that "pre-emptive wars" such as the Iraq War are morally justified, we find this attempt resisted by the fact that, contrary to the opaque intelligence claims of our trusted leaders, Iraq neither possessed nuclear weapons nor was involved in the 9/11 terrorist attacks. We are then forced to rethink the entire process through which the notion of "pre-emptive war" *inevitably allows behind-the-scenes players* to make the determination as to whether pre-emption is necessary in any given case. At this point, we realize that "pre-emptive war" is not as simple a notion as we at first may have thought. A vast array of real-life facts and questions "resist" our pretenses of knowing what the right action-guiding policy might be.

Here again, we know we are dealing with a reality and not with an arbitrary subjective whim because of the way reality resists our attempted actions – in this case, the attempt to form a coherent belief about the world. Although Panksepp seldom uses the term "enactive" or "enactivism," it is clear that the role of the SEEKING system is to lead the organism to act in self-energized ways that are not merely caused by environmental inputs, but rather are self-organizational. Emotions don't simply transform stimulus inputs into organismic responses. Instead, the organism is acting *upon* its environment to achieve valued self-organizational and action-oriented goals, within the larger self-organizational structure of a social environment.

Action, as opposed to mere reaction, requires a self-organizing structure. Self-organizational systems are systems with a built-in structural tendency to maintain a continuity of *function* while appropriating, reorganizing, and replacing the micro-constituents needed to keep the patterns of activity going that define the system's identity. It is then a short step to the "enactive" theories of consciousness and emotion proposed by Shaun Gallagher (2006), J. J. Gibson (1986), Natika Newton (1996, 2000), and tracing back at least as far as Maurice Merleau-Ponty (1945, 1942). We understand our environment largely by acting in relation to it, and we understand other people not only in this same way, but also by imagining what it would be like to perform *their* actions (Newton 1996, esp. ch. 6).

Correlatively, we know that external constraints are *real* precisely because they constrain our ability to act. Thus the very difficulty of forming a coherent moral worldview is an indication that the principles involved are not merely arbitrary whims on our part. We wish we could just believe whatever we want in the moral realm, but our own need for epistemological coherence resists our attempts to do so.

1.9 Moral Development Requires a Striving for Epistemic Coherence

Some may be inclined to think that natural altruistic feelings are adequate to motivate a "good enough" morality. Why can't ethics just consist of using "wisdom" or "self-awareness" in exercising our already-existing empathic inclinations?

The problem is that this view precludes forming non-trivial ethical judgments. A simple emotivism merely advises us that we should do whatever we are motivated to do. If we are religiously intolerant, and if there is no apparent or strong enough motivation to stop being intolerant, then on the strictly emotivist view there can be no corresponding ethical injunction. It is simply a descriptive fact that many people have been motivated to be religiously intolerant – and people can't act contrary to their motivations, at least not within a naturalistic understanding of psychology. The problem still remains that, if not for an independent, unconditioned desire to know the truth, we would believe whatever was most hedonically "reinforcing" to believe.

The exploratory drive, by contrast, leads creatures with considerable intellectual capacity (for example, humans) to seek out the most *coherent achievable account* of what kinds of things have value. One of my central contentions is that in this process we learn that we can't coherently posit that *creatures who value things* do not also have value themselves. This is very similar to the definition of "value" used by personalists (for example, Randall Auxier 1995; Edgar Brightman 1933; Thomas Buford 2011). Our challenge then is to construct the most coherent possible moral worldview to accommodate this value. The appreciation of the valuational importance of others qua valuing creatures, along with the perception of their vulnerability as such and the pathos of their actual or potential suffering, then inspires us to action.

A still more interesting question in this regard is how we justify behaving as if some valuing creatures do *not* have as much value as others – members of other races, political enemies, people with unpleasant personalities, "lazy, irresponsible bums," and generally everyone whom we manage to vilify for one reason or another, not to mention those whom we ignore altogether. It is very typical, for example, to believe that "immoral" people don't "deserve" moral consideration for their suffering, even if their bad acts are continuations of a cycle of abuse or the results of abnormal brain function such as schizophrenia or bipolar disorder. How do we coherently persuade ourselves that various classes of valuing creatures somehow inherently "deserve" to receive more suffering and misery than is necessary for deterrence purposes, or when such suffering doesn't serve to create any *beneficial outcome* for anyone?

In a recent botched execution, the drugs failed to work properly and the condemned man struggled and writhed around for almost two hours gasping for breath before finally dying of suffocation. Family members of the condemned man's two victims said afterward that the public spectacle of this torture didn't bother them in the least, because the man "deserved" to suffer. What are the coherence criteria for making these distinctions with regard to which valuing creatures' well-being or suffering does or doesn't have at least prima facie value? If no good came of the man's suffering, then what is the countervailing prima facie value that justifies his suffering, and thus the negation of his status purely as a sentient creature, even if a morally corrupt one? Opinions about such things can be completely unfounded, or they can be epistemically coherent to greater or lesser degrees.

By speaking in terms of people's interests having value "all else being equal" or "prima facie," we can avoid being detained by the obvious fact that people often value things that are harmful or bad in the final analysis. Conflicting prima facie values have to be balanced and weighed against each other. The positive prima facie value of a heroin addict's pleasure is more than counterbalanced by the amount of *negative* value caused by becoming a heroin addict. In the same way, we balance the freedom of speech against the harm of advocating assassination of a political opponent.

So the idea that all valuing creatures have prima facie value doesn't automatically imply that I should devote the same time and energy into pursuing all people's well-being, or that I should not pay special attention to my family and friends, or that I should contribute 99 percent of my income to charities serving various needy people around the world. The idea that all valuing creatures have prima facie value doesn't even necessarily mean that, in the final analysis, I should not regrettably harm others when needed for a greater good – as when it is necessary to punish criminals in order to deter crime, or to tackle another player when playing football.

The question as to who is responsible for whose well-being, and what is the best way to distribute our efforts, requires working out an entire coherent value system, including how we believe a social contract or community of persons should work, and what the criteria should be for assuming different kinds of responsibilities. Most such systems will allow that, at the immediate present moment, I am more responsible for focusing on the soldier next to me during combat than I am for thinking about the well-being of children in far-away orphanages. On the other hand, after combat is over, it may be incoherent to continue feeling as if only the interests of my erstwhile comrades-in-arms call for an important moral commitment.

We still have not yet broached a number of important issues involving internal psychological conflict when it comes to the attempt to establish and maintain a coherent moral worldview. Here is a breakdown of the main topics of the remaining chapters of the book:

Chapter 2 summarizes a few relevant aspects of the neuropsychology of the endogenous emotion systems insofar as these various systems contribute to understanding ethical thinking and behavior – especially the exploratory drive and Panksepp's SEEKING system. We then begin to touch on the psychology of the "darker side of irrationality" as it comes into conflict with the truth-seeking motive. In this context, we also look at the implications of the new emotion psychology for the "naturalistic fallacy" problem in ethics, which has been a major stumbling block for both moral theory and actual moral belief systems throughout at least the past century.

Chapter 3 then explores the way *selective failure* of the exploratory drive can lead to moral incoherence and confabulatory belief systems. We especially examine the basic inner conflicts between the exploratory drive and other emotional purposes that lead us toward selective suppression of the exploratory drive, resulting in distortions of thinking as illustrated by Theodor Adorno's work on "authoritarianism" and the Ernest Becker/Jeff Greenberg "terror management theory."

Chapter 4 begins to explore ways in which emotional psychology affects how we think about the epistemology of ethical belief systems. The main focus of this chapter is to understand how a *subjective* act of "valuing" can relate to a system of truth-functional *beliefs* about morality. It is necessary to clarify differing but equally legitimate meanings of value-laden terms, with a view to understanding why people need to integrate all of our various valuations into a coherent system. We examine a crucial argument for the natural impulse to regard value as something that really exists independently of ourselves, rather than merely an expression of our own emotions. The overall argument of the book is briefly summarized in this chapter, including why a coherent value system will *universalize* our valuing of others because we intrinsically value the act of value-creation on the part of any and all valuing creatures, and correlatively we negatively value the suffering that comes with being a valuing creature per se, at least prima facie.

Chapter 5 then lays the groundwork for an understanding of ethical thinking and behavior in terms of a "coherence" rather than a "foundational" epistemology – with "coherence" understood in a sense that would increase the probability that an overall moral worldview is more likely to be true than other alternative worldviews, all things considered. In this context, we examine the role of motivated *selective attention* as it affects our valuing or non-valuing of different people and their well-being. We explore some of the motivations that distort our selective attention, as well as the way selective attention affects the coherence of a moral worldview.

Chapter 6 examines the crisis of contemporary ethical thinking, in academic philosophy and psychology as well as everyday life. The problem of the *"hermeneutic circle"* becomes central in this analysis: The hermeneutic circle stems from the fact that our love of truth *requires* that we engage in an inevitable process of interpreting each new experience in such a way as to try to fit it to an already existing worldview (hence the term "hermeneutics" or "interpretation"). Since we can't know for sure how much this distorts our view of reality, and thus how severely we ourselves are trapped in the self-imposed distortions of a hermeneutic circle, dealing with this dilemma requires a kind of coherence epistemology that allows a place not only for empirical psychology but also for careful phenomenological investigations of affective-emotional processes that affect our interpretation of reality, granting that such interpretations will always be fallible and subject to further hermeneutic questioning. At the same time, we want to avoid elevating phenomenological introspection into the kind of "moral intuition" that so often serves only to prop up the prejudices of a hermeneutic circle once it is in play.

For example, the "terror management theory" inspired by the work of Ernest Becker (1973), and later pursued by Jeff Greenberg et al. (1997), is largely a psychological investigation of what it is like to be trapped in an extreme case of the hermeneutic circle. Certain inadequate attempts at "terror management" can lead to something like what Adorno et al. (1964) and Altemeyer (2008) call an "authoritarian" worldview (a concept that I admit needs clarification and refinement), as the next-best semi-coherent worldview that some existentially terrified subjects are able to achieve. From Becker's perspective, people fall into such worldviews because their terror (in the face of inevitable human emotional problems, not just those involving physical survival) forces ignoring of some of the facts that would be relevant if more coherent worldviews were to be formed. It is difficult to admit to ourselves many terrifying facts, including the real extent of our relative powerlessness and unimportance in the ultimate scheme of things, not to mention the inevitable shortness of our life span and the life span of our loved ones.

But even in the optimal case, the best we can do is to evaluate the coherence of overall views of reality into which certain statements either can or can't be fit. Whether we accept or reject a given statement often depends on how well it fits into an already existing overall worldview. In the quest for coherence, we can either adjust our overall worldview, or we can reinterpret the facts to try to make them fit the existing worldview. Sometimes it makes more sense to hold onto the overall worldview while searching for ways to resolve contradictions created by apparently anomalous experiences. Other times we hold onto the previous worldview not because it makes more sense to do so, but because we feel threatened by the change.

Chapter 7 argues for the coherentist structure of moral thinking by considering both psychological and philosophical assumptions of several failed yet highly instructive examples in which there is confusion as to whether the arguments intend to foundationally "prove" ethical propositions. There are both positive and negative lessons to be drawn from these endeavors. Logically based ethical thinkers like Kant, Gewirth, and Rawls are precursors to what I call "coherence epistemology" in ethics, because none of them pretends to deductively "prove" ethical propositions without substantial initial assumptions, especially assumptions about our emotional dynamics. On the other hand they do assign a vastly predominant role for deductive logic in establishing what they consider, in essence, to be ethically binding norms.

Despite these thinkers' over-emphasis on deductive logic, their focus on *emotional* issues such as the involuntary "humiliation" of the moral conscience (Kant 1788/2002) and the need for "agency" (Gewirth 1980) can be illuminating. Correlatively, the modesty of these philosophers in not claiming apodictic certainty for their arguments allows us to learn a good bit from them in the attempt to ground ethics in a coherence rather than foundationalist epistemology.

Chapter 8 consolidates the case that what is necessary – and in real life what we actually use – is a "big tent coherence" epistemological approach. By "big tent," I mean an approach that includes a variety of epistemological tools, including careful introspection into the emotional processes that ultimately affect the shape of our hermeneutical interpretation of reality. This approach is very different from a method that would emphasize foundationalist "proofs" of moral principles.

The foundationalist approach remains inadequate even where it tries to use *empirical* evidence in addition to deductive logic in grounding its claims. In my view, the crisis of moral epistemology stems from the fact that, just as in natural science, we can't conclusively prove the truth of ethical statements based on the foundation of some unquestionable first premises – nor can we unequivocally deduce them even from some probabilistic empirical premises or other. But in science, no one suggests that the lack of certainty means that nothing can be known with some degree of fallibility, nor does anyone suggest that scientific theories are "only expressions of emotion."

Chapter 9 fleshes out criteria for a workable coherence approach, emphasizing the effects of our conflicting emotions on the relative incoherence of our overall moral worldviews. Thomas Kuhn (1964) and other philosophers of science point out that science itself exhibits what we have been calling a hermeneutical structure, in which facts are fitted to theories as well as the reverse. But if the same is true for ethics, I want to argue that this situation should no more entail a relativist or emotivist position than it does in science. Just as in science, there are epistemological elements that can establish that some value systems are better than others. This is what I mean by a "coherence" as opposed to a foundationalist approach. There are epistemological principles for ethical thinking that can discriminate between more and less coherent moral worldviews. But again, these principles must include – and in our everyday thinking they often do include – much more than a mere avoidance of logical self-contradiction. In everyday life, we usually have what Gendlin calls a "felt sense" of what we should be doing in each instant, relative to short-, medium-, and long-term values; but when the inspiration to short-term action fails, we are reluctant to connect this failure to a problem with our more *long-term, intrinsic* values. It becomes necessary in this context to place the resulting implications for valuational thinking into a larger context by examining the inevitable continuity and commensurability of "moral" values with "non-moral" values. When we periodically allow ourselves to question the medium- and long-term values toward which the short-term ones are instrumental, the self-organizational system can continuously renew the balance of various values, which, again, are always present in an inherently acting creature.

The Conclusions chapter then pulls together all of the forgoing discussions into a unified theoretical approach. But first things first. A psychological prolegomenon to any coherent approach to ethics is to reconcile morally relevant natural emotions with any non-trivial ethical statements beyond "Act morally when and if you are naturally motivated to do so" (or, more simply, "always do whatever you want"). The next chapter will focus in a little more detail on the recent neuropsychology relevant to this problem, and the philosophical implications of the new ways of thinking about moral psychology.

NOTES

1 That some instinctual behavior patterns don't need to be instrumentally reinforced by an external "reward" is now fairly non-controversial. What remains more controversial is whether such internally motivated behaviors are merely "rewarding" in themselves, without external reinforcement, or on the contrary whether the behaviors are executed even in instances where they are *not* particularly pleasant or enjoyable – as for example Martha Stout (2005) believes with regard to the motivating power of the "moral conscience." This question will be addressed more fully in our Chapters 2 and 3.

2 Phillip Luce (1960) made a detailed study of the notorious racist organization called "The White Citizens Councils" and wrote: "The attempt to enlist the better elements of the community goes along with the philosophy of the Council that [quoting here from the Council's newspaper, *The Citizens Council*] 'its members wear business suits rather than bedsheets.'" In Jackson, Mississippi the members are the "pillars of the community. They are the kind that run the Chamber of Commerce and the Community Chest, serve as officers of churches and do civic chores in every town worthy of the name" (Luce, p. 35). Luce (esp. pp. 35 ff.) looks into specific individuals involved and finds predominantly that they indeed were heavily involved in charitable and church work, and were public-spirited community leaders in their towns, besides paradoxically working toward extremely oppressive white racist causes, ostensibly extending even to involvement in the murder of African Americans and subsequent legal cover-ups, such as in the murder of Emmett Till, who made the mistake of looking at a Southern white woman in the wrong way. The *New York Times* (February 12, 1957), commenting on White Citizens Councils, wrote that "Because their rolls read like a Southern 'Who's Who,' a halo of respectability hovers over them."

3 Moreover, Gilligan and Cornell argue, Kohlberg's test of "moral maturity" presupposes that Kohlberg's own flawed understanding of moral philosophy is the correct one. But Kohlberg's moral system glaringly fails to deal with the "why be moral?" problem in any coherent way. Hence his assessment that boys are more morally mature than girls, even though they fail to *act* in a more morally mature way.

4 The relationship between "motivation" and "emotion" varies from author to author. Damasio speaks of "emotions" as present physiologically, independently of consciousness of them. Others speak of "emotion" as the conscious feeling component *of* "motivation." For Panksepp, every emotion system is also a motivational system; emotions are always potentially capable of leading to conscious feelings. Interpreting the "meaning" of an emotional feeling – what it "is about" – is a difficult and error-prone process. For now, we use

"basic emotion systems" in Panksepp's sense – as always including "motivation" systems.

5 For example, it is now beginning to appear likely, through the work of Mayberg et al. (2005), Fox et al. (2012), and other recent neurophysiologists, that Bartleby's overactive Brodmann's area 25 in his anterior cingulate had probably in turn shut down other parts of the anterior cingulate and thus severely dampened the entire SEEKING system. But the physical details are not so important for our purposes here.

6 Notice also that one of the main neurotransmitters that fuels the SEEKING system is dopamine, as will be discussed in the next chapter. Dopamine is always involved when we perform actions.

7 In addition to depression exacerbated by dopamine deficiency, depression can also involve serotonin deficiency (Tremblay et al. 2005); serotonin is a more complicated neurotransmitter that works mainly by regulating the balance of other neurotransmitters. In that case, the causation is indirect, and normal physical energy may occur but not be usable toward a goal felt as worthwhile. Thus many psychiatrists (e.g., Dunlop and Demeroff 2007) warn that antidepressant drugs that simply elevate dopamine can increase the likelihood of suicide. The patient may lack a feeling that living is worthwhile, yet now have the energy to carry out a suicide.

8 Many psychologists refer to this condition as one of "anhedonia." But Gabbay et al. (2013) show that anhedonia is only one component of depression. It is not merely the inability to feel pleasure that characterizes depression, but rather more extremely the inability to feel the value of the outcomes that, *if we did* feel their value strongly enough, could give us a sense of satisfaction when achieved.

9 Heidegger and Jonas observe that we have no a priori clear understanding of *what* those future potentialities are that drive our "caring." We have "instrumental" values before we ask ourselves what are the ultimate ends toward which the instrumental values *are* instrumental. (Thus for Heidegger we begin life in "inauthenticity.")

10 Granted, twentieth-century psychology recognized not only reinforcement learning, but also certain *"instincts."* But the instincts relevant to cooperative and helpful behavior were usually explained by evolutionary bioethicists as methods of preserving the self-interest of the species as a whole (Dawkins's "selfish gene"). And even without this assumption, the problem remains that "Act morally when and if you feel like it" still doesn't seem to have any nontrivial ethical content.

CHAPTER 2

Can We Have Naturalism without the Naturalistic Fallacy?

It has become fashionable to dismiss our old friend the naturalistic fallacy as irrelevant. Dennett, for instance, in his earlier work (1984) wrote off the naturalistic fallacy problem as essentially a mere intellectual parlor game. Who cares, he questioned, whether we call a naturally motivated altruistic action "moral" or not? The difference is merely a quibbling over language.

But then in his later work on naturalistic ethics (2003) Dennett saw that it was necessary to take the problem seriously. Even in this later analysis, however, the reason the naturalistic fallacy becomes problematic is very narrow – it is only because of the problem of determinism: If people's behavior is predetermined, how can we hold them responsible except to do just whatever they were predetermined to do?

The determinism issue is not a deal-breaker for the naturalistic fallacy problem. To be sure, it enters into the choice as to which moral viewpoints are more coherent than others, but it doesn't render them all incoherent. If someone refuses to pay back a loan, the lender often may have a right to force repayment regardless of whether the borrower was predetermined to be the kind of person who would refuse to pay back the loan. Similarly, even though a criminal's personality etiology may be predetermined, if the criminal refuses to fulfill his obligation to obey the law, others often may need to impose punishment if that is what is necessary to deter crimes in the future – even if they also regret having to harm the offending individual. We may have a right to punish criminals even if they can legitimately claim, as Bill Faw (2000a) puts it, "My amygdala-orbitofrontal circuit made me do it!"

But a much thornier problem than determinism is the one that concerned G. E. Moore at the very beginning of the twentieth century (Moore 1900). Avoiding the naturalistic fallacy is crucially important if we are to avoid effectually *trivializing* ethics. The role of ethics in our lives can't be served by reducing it to a series of empirical descriptions of the way various people happen to feel. We saw in the previous chapter that a prescription that we should act altruistically when and if we happen to feel altruistic provides no moral guidance at all. Yet recent attempts to reconcile ethical theory with the human sciences have tended to focus almost exclusively on emotions that lead directly to compassion, empathy, or cooperation – in short, to altruistic emotions that we already happen to feel, "by nature" (for example, Bowles and Gintis 2013; Bräten 2007; Filonowicz 2008; Roberts 2013).

Suppose the following statements are true:

(1) I ought to enhance Bill's well-being, and I also happen to want to.
(2) I ought to enhance Mary's well-being, but I don't want to.
(3) There is no moral reason to enhance John's well-being, but I want to.

If I can do only what I am motivated to do, the result is that I will enhance *Bill's and John's* well-being, but *not Mary's*. And this is exactly what I *would have done* without any ethical injunctions. So if this were the best we could do, then ethics would still reduce to "Always do whatever you already would have done without any ethical injunctions." And this, again, is tantamount to "Always do whatever you want *simpliciter.*"

Someone might think that there are quick or easy ways to avoid this problem. For example, we might try emphasizing the ambiguity of "can" as it pertains to the "ought implies can" principle. Perhaps we "can" be motivated by things we are not yet motivated by, if we can *learn* to be motivated by those things. Obviously, as Dennett (2003) among others points out, the "can" in "ought implies can" may mean very different things. Currently, I "cannot" run a marathon, but with the right training (which I know how to do, or at least *can find out* how to do), I *could* run the marathon, so in that sense I "can." Similarly, the American racists of the "Old South" – unlike the cat chasing the mouse – "could" have had different psychological attitudes if they had been reared differently or if they had made different choices earlier in life.

But these earlier choices *also* would need to have been motivated. So the question remains whether those earlier choices *could* have been made given the circumstances at the time. Is it realistic to expect that an Old South segregationist "could" have been motivated to question the entire fabric of Southern American culture, unless some accidental trigger happened to motivate the person to do so? The ambiguity of "can" in "ought implies can" doesn't deliver us from the problem that moral action must be motivated; yet moral principles risk being trivialized if our only motivation for them is a natural inclination to engage in certain altruistic kinds of behaviors.[1]

In short, altruistic and cooperative feelings, whether innate or learned, are not enough to establish non-trivial ethical judgments. If it were not for the independence of the SEEKING system and resulting truth-seeking motive, then the end game of moral inquiry would already be predetermined by whatever emotional preferences I already happen to have.

Luckily, as this chapter will explore further, emotional preferences *don't* completely dictate what we believe, because the exploratory drive doesn't merely want to discover *that* the truth is XYZ due to some emotional preference for XYZ. Instead, it wants to discover *what* the truth is, regardless of what our competing emotional preferences might be. Or to put it the other way around, the exploratory drive dictates *that* we seek the truth, but doesn't dictate *what* we take the truth to be.

Since this notion of an independent exploratory drive plays such a crucial role in allowing escape from the naturalistic fallacy, and from the tyranny of simple hedonism, it will be helpful to explore the neuropsychology of this situation in a little more detail. The next section shows why recent neuropsychology supports Hume's idea of an independent "love of truth" as a basic tendency of the mammalian brain, resulting especially in abstract truth-seeking in humans.

2.1 Basic Neuropsychology of the Truth-Seeking Motivation

Panksepp's physiological studies suggest that, if we count the system that he calls the PLEASURE/PAIN system, there are eight relatively independent, endogenous, and unconditioned motivational/emotional brain systems. While not exhaustive from an experiential point of view, these most "basic" systems that Panksepp studies are the most *innate and hardwired* of emotion systems – the ones we are born with, and whose neurophysiology will always remain with us throughout our lives, however much more complex and subtle our affective feelings become as we interact with the environment and especially the social environment. The next section will further explain why I think it is so important for our purposes to take note of these "basic" emotion systems. Yet the eight emotion systems in this list are *not* necessarily meant to describe or exhaust the vast array of everyday human phenomenally felt affective feelings, which can vary extensively, sometimes even in extreme ways. (Notably, moral *guilt* is one of those more complex emotional processes that Panksepp does *not* reduce to one of the basic brain systems. More on guilt later.)

Part of the reason we know these different systems are independent of each other and not derivative through reinforcement is that each is subserved by its own specific brain pathways and neurotransmitter systems, none of whose structure or function depends on any of the others through reinforcement or learning, and all of which develop from innate growth and development patterns. In experiments and neurological studies involving lesions of one or another of the systems, we find that the non-affected systems continue functioning relatively normally and don't depend on reinforcements derived from other systems. The different emotion systems are all physiological tendencies of the organism that don't need to be externally or instrumentally rewarded in order to express themselves. They occur *without conditioning.* Similar conclusions are reached when a particular system is stimulated or suppressed through artificial chemical enhancement of the neurotransmitters involved. For example, excess dopamine hyperactivates parts of the SEEKING system, resulting in behaviors typical of this system; or hyperactivation of one part of the system can inhibit other parts (see below). The work of Doug Watt (2000) also supports Panksepp's contention on the endogenous nature of these emotion systems.

I like to break Panksepp's analysis into eight relatively independent systems, which are described in more detail in Appendix 2.1 (see pp. 68–70). Here is a listing of them:

1. The SEEKING system (which forms the neurophysiological basis for what many call the "exploratory drive"). This system activates when we focus *voluntary attention*, initiate *action as opposed to reaction*, become cortically aroused to *energize action*, or feel *curious or interested.* And as noted earlier, its activity is necessary to avoid clinical depression.
2. The RAGE system includes the much-discussed medial amygdala (the "old amygdala" mentioned earlier), but it also includes much more extensive brain areas described in Appendix 2.1. As an example of how independent the different systems are, RAGE includes the *medial* hypothalamus (rather than the *lateral* hypothalamus emphasized in the SEEKING system). When people speak of a "fight or flight" response, they are actually conflating two brain systems, RAGE and FEAR, although there is more substantial overlap between these two systems than others.
3. The FEAR system, as most people now realize, includes parts of the amygdala (central and lateral amygdala – again, primarily the "old amygdala"), but also many other brain areas that differ from the other emotion systems. One of the neurotransmitters it uses is the notorious "corticotrophin releasing factor" (CRF) that increases with stress and eventually can cause heart and circulatory problems. Physical exercise

tends to reduce those medical problems by "burning off" the excess CRF, which if not burned off could lead to further stress and a vicious cycle of increasing CRF. So we can see the connection of this system to the everyday experiences of anxiety, stress, fear, and worry.

4. The LUST system is a widely distributed system that runs from deep in the subcortex (the PAG – see Appendix 2.1) all the way to the hypothalamus. Among the neurotransmitters it uses is the much-discussed *oxytocin*, a substance that contributes to loving feelings; oxytocin increases in the brains of both men and women after having sex.
5. The CARE (nurturance) system runs from the periaqueductal gray (PAG) of the deep subcortex, through various brain areas (see Appendix 2.1), all the way to the anterior cingulate (which is active in the direction of attention) and some prefrontal areas. This system seems to include both the nurturing and nurture-needing experiences and behaviors of dependent children and care givers. The dramatic *increase* of oxytocin *during pregnancy* is a well-known example of the way this system affects nurturant feelings. Women for whom abuse of certain drugs such as cocaine has interfered with oxytocin chemistry tend to bond less easily with their newborn infants. Among mammals, females have much more oxytocin than males, but the males do have some oxytocin and also compensate with similar effects of prolactin and vasopressin.[2]
6. The PANIC/SEPARATION DISTRESS (social bonding) system includes brain stem areas and runs all the way up to the thalamus and preoptic hypothalamus. Panksepp uses the term "PANIC" as distinct from FEAR, because here we are speaking of a special system for *social bonding* emotions, which when disturbed produces the intense PANIC reaction, especially in infants separated from their caretakers and vice versa. It is very different from the FEAR reaction to a stimulus perceived as threatening.
7. The PLAY system, running from midbrain central gray to posterior thalamus, relies extensively on opioids and acetylcholine. This is the system activated in Panksepp's famous studies of rats who, when playing, exhibit "rat laughter" (described, among other places, in his 1998 book). This system is very distinct from the SEEKING dopamine circuits and the circuits in other motivational systems.[3]
8. The PLEASURE/PAIN system runs from PAG to medial hypothalamus (not the *lateral* hypothalamus, which is active with SEEKING) – relying more extensively on opioids, as regulated by GABA, than do the non-consummatory tendencies of the SEEKING and PLAY systems.

Those who are knowledgeable about neurophysiology will be interested in seeing more of the details (see Appendix to this chapter) demonstrating that, in fact, the different emotion systems are subserved by different combinations of brain areas and different neurotransmitters in different interactions with each other.

The crucial thing to notice here for our purposes is not so much the nitty-gritty of the physiological details, but generally how distinct the different brain systems are from each other (although there is some overlap between the RAGE and FEAR systems, and also between the LUST and CARE/NURTURANCE systems). We can see, even from this simplified summary, that each system uses a different combination of brain areas and neurotransmitters. Remember that the all-caps nomenclature indicates that Panksepp is not using these words in their everyday meanings. Notably, CARE refers not to moral and social emotions generally, but specifically to nurturance, as in a parent's caring for offspring.

Counterintuitively, PANIC refers at least as much to the social emotions as does CARE, because the model for "PANIC" is the infant's distress when separated from caretakers and siblings. Conversely, the paradigm of "CARE" is the parent's tendency to care for the young. The PANIC system would be triggered in adults when a separation or divorce occurs between lovers or spouses, or in grieving for a deceased loved one, or during feelings of extreme social isolation. Hannah Arendt (1963) emphasizes the role of social isolation feelings (sometimes called *"anomie"*) in leading to extremely irrational and even totalitarian social belief systems.

Panksepp's analysis of the SEEKING system is consistent in some ways with Descartes' view that "wonder" or "curiosity" (Descartes' *estonnement*) is the most basic of all the emotions. In Descartes' view, it determines the direction of attention; it therefore leads to the search for knowledge or awareness of any other object or situation toward which *any other* emotion might be felt (Descartes 1649/1989, pp. 56–58). Similarly, the SEEKING system, rooted in the deep subcortex, must activate if we are to be energized for any self-initiated action, including the voluntary control of the direction of attention.

The crucial point for our purposes is that, regardless of our environmental experiences, we each come into the world with a number of basic emotional and motivational tendencies that aren't learned and don't need to be reinforced with rewards or punishments in order to manifest themselves. None are derived through learning from reinforcement of the others. On the contrary, each one is endogenous in the brains of mammals, and is active prior to and even despite any reinforcement. Granted, there are pushback arguments from behaviorists against Panksepp's way of thinking, and those will be discussed briefly in the next chapter.

Most important for our purposes, the exploratory tendency (an aspect of the "SEEKING system") is an unconditioned motivational system, not acquired through reinforcement of other drives. It therefore independently motivates us to know the truth. We should also be aware that different systems can mutually inhibit each other. For our purposes, the take-home message is that the different motivations are not derivative through reinforcement, but instead are primary. This issue will be discussed further in the next chapter. But first, let's begin to consider the implications for moral psychology.

2.2 Why Focus on "Primary" Emotions?

Why should we focus on "primary" (or "basic," "innate," "unconditioned") emotions like nurturance, social bonding, or the exploratory drive – rather than, say, the nuanced and sophisticated affective response to poetry or music, or the concrete feelings of a particular soldier while in combat? I don't by any means want to suggest that human affective life is reducible to or completely determined by primary physiological drives. The most interesting and important aspects of human affectivity are the ones that are *not* orchestrated in such a simple way. Far from universal, those are unique and subtly variable through literally infinite possibilities.

The reason the naturalistic fallacy problem requires focusing on primary drives – the ones we are born with and can't escape – is precisely for that very reason. The primary emotions can't be disowned or denied, because they are demonstrably *necessary* aspects of the human affective situation, although they can be tweaked in an endless variety of ways. All humans are partially driven by our primary motivations. No one can disown or escape them, as perhaps we could claim to disown sympathy for the poor by blaming them for their own circumstances.

Granted, if I were ideally in touch with my inner life, I could perhaps see that blaming the poor for their own problems is not what I "really" feel. But in the context of public discussion and philosophical thinking, subjective self-reports are suspect; anyone can claim to feel whatever suits their hermeneutical purposes, or misrepresent their feelings *even to themselves*.[4]

What we *can't* disown are the emotional processes that are inevitable parts of being human. We can't say that we don't happen to have any altruistic feelings per se (initially, toward specific individuals and in particular situations, again with the qualified exception of extreme and rare sociopaths). In the same way, we can't escape the fact that an exploratory drive motivates a search for truth, independently of whether believing in such truths gratifies our other desires. We can't disown those endogenous motivational systems because they are a non-controversial part of the makeup of our organisms, demonstrable in a hundred different scientific ways, as well as through phenomenological experiencing.

The same is true for the exploratory drive. Such a basic drive can't be written off as if it were an idle whim, as in Ayer's "I don't like murder" – as if such a sentiment could be as arbitrarily variable as "I don't like ice cream." Later I will show that these primary motivational systems, when acting in combination, prevent us from denying that we experience a "moral conscience," other than the 1–4 percent who are complete sociopaths. As Stout (2005) summarizes the empirical research (she is explicitly in agreement with Kant on this point), the presence or absence of a conscience *cannot be primarily attributed* to the vicissitudes of socialization, although it may be tweaked by them. Even many vicious criminals are not true sociopaths, because they do experience moral guilt about some things, while rationalizing away their guilt feelings about other things (Anderson et al. 2009; Faw 2000a, 2000b, 2003; Koenigs and Tranel 2006). To be sure, brain development can also be influenced by environmental conditions. But we have already seen the strong role of biological brain growth as a consistent factor affecting moral development; and of course this growth process also can be crucially affected by early blows to the head.

I am not arguing that the inescapability of certain emotions such as "care," "social bonding," or even "feelings of conscience" by themselves can ground moral beliefs in a non-trivial sense. We have already seen why they can't. But when we combine those emotions with the independence of the "exploratory drive" aspect of the SEEKING system, we then end up with a motivation to pursue the truth about what our value priorities should be. Emotions such as "care" and "social bonding" provide the first building blocks for a concept of "value," but it is the exploratory drive that leads us to formulate a coherent accounting of our total system of value priorities.

The most interesting focus for our purposes here is that we can explore moral thinking and moral conflict from a new perspective. The basic desire to know the truth conflicts with the ever-present temptation to believe what makes us feel good, for example by propping up a pre-established worldview. Neuropsychologists now mostly agree that there are innate tendencies to value specific individuals in our family, clan, or circle of friends;

this results from elaborations and developments of the social bonding systems in Panksepp's list. But it is the exploratory drive that forces us to ask ourselves whether this same value may also extend to all similarly valuing creatures, even when it would be more pleasant *not* to extend empathic sentiments toward those with whom we are in conflict. Any such moral truths that may be discovered thanks to the exploratory tendency can also conflict against feelings of social acceptance from an "in-group," which salves the PANIC of social isolation, and yet may demand *lack* of empathy with "out-groups."

Similarly, it would be more pleasant not to extend pro-social feelings toward those for whom compassion would demand paying a certain price, even if small (for example, the cost of helping the poor) – or for that matter toward those who themselves are deemed immoral and thus with whom it can be difficult to empathize. Does the truth require a certain kind of coherence in the way we value the well-being of various others? Does it affect how we balance competing values, and whether we universalize our valuing of others? It may be easier simply to direct attention away from the small farmer down the road who lacks health insurance for his wife's cancer, than to confront the overwhelming question as to what should be done. Yet our exploratory drive, including the "love of truth," motivates attention to these kinds of questions.

Questions like these would come naturally to creatures with our intelligence, except that there are also conflicting motivations to deflect attention from them. The next section begins to explore some of those conflicting motives.

2.3 Is Moral Thinking "Fundamentally Irrational"?

Recent work in moral psychology has increasingly emphasized the irrationality of the emotions that lead to ethical actions or belief systems (for example, Hibbing et al. 2013; Lakoff 2008; Westen 2008). People don't think in terms of deductive and inductive logic, say the intellectuals (who, by contrast, *do* think logically). Instead, we are told that "people" (other than the intellectuals making the assessment) use only "imagery" and "framing narratives," at least when it comes to moral and sociopolitical issues. Despite the fact that the same ordinary people do an admirably logical job of keeping our computers working, our cars running, and our students registered into the right classes, they are supposed to suddenly become incapable of rationality when it comes to social and ethical issues.

Some might cite the 2016 US Presidential election as evidence of the extreme irrationality of the way people make value decisions (and thus sociopolitical ones). But we should also remember that the 2016 election cycle illustrates the failure of both the politicians and the media to *offer* people rational arguments to support the claims they were making about what the best social and political policies might be. Instead, they mostly asserted their opinions without supporting arguments, and exchanged insults about each other's character. Even when rational arguments were presented by the candidates, those went mostly unreported in the media, unless someone were to exert a heroic effort to find them. Nor was there much public discussion of the rational arguments for and against those policy opinions.

Both the politicians and the media *had already assumed* that people are not capable of basic logic, and therefore didn't present arguments to support the assertions being offered – thus leaving the people to flail around in the resulting irrationality. By contrast, during the 2012 re-election of President Obama, Obama was in danger of losing the election until Bill Clinton at the Democratic National Convention clearly communicated a logical argument against "trickle-down" economics as a strategy for stimulating economic growth and job creation (see New York Times 2012). He pointed out that workers are also consumers; if workers have little money to spend, companies can't expand production because the market for their goods and services is depressed, no matter how many tax breaks or deregulatory policies are offered to the companies. There simply is not enough demand for their goods and services to warrant expansion, because the worker-consumers are not earning enough money to buy them. Many people understood this argument. The Democrats got a huge "bump" in the polls after that convention.

Obviously, people do understand simple logic when motivated to focus attention. If people were incapable of logical thinking, we would have no computers or nuclear submarines. Someone had to be intelligent enough to invent, build, maintain, and operate them. The question is whether the same capacity for rationality can be applied to valuational belief systems.

Irrationality, if it were to completely permeate the morally relevant emotions, not only would eliminate any truth-functional approach to ethics, reducing it instead to an empirical psychology of moral sentiments. A worse consequence would be that it would also render impossible any solution to the "anything goes" tendency in moral epistemology, because there would no longer be anything more to say about moral sentiments than "Well, that's just the way I feel."

If moral action were motivated only by emotional preferences – even altruistic ones – then the emotivists would be right: "Murder is wrong" would mean nothing more than "I don't like murder" – which I am suggesting is too simplistic to serve as an adequate understanding of how moral emotions or even moral epistemology work. Nor can it serve as a practically useful guide to actual ethical decisions. Jean-Paul Sartre would be correct in his assertion, at the end of *Being and Nothingness*, that "It makes no difference whether I get drunk alone or become a leader of nations" (Sartre 1943/1965, p. 627). By contrast, the independence of the exploratory drive motivating a "love of truth" opens up new possibilities.

It might be argued that moral questions are fundamentally different from questions about computers and submarines. Value issues might be irrational in ways not applicable to those involving the physical world. The question, then, is whether some of the same capacities for evaluating beliefs are applicable to all types of beliefs, and whether similar psychology is relevant to the formation of belief systems in both the valuational and the empirical domains.

We must also consider how our motivation to seek the truth affects the *ways in which* we do so. For example, it will become obvious that truth-seeking inevitably always presupposes a previously existing overall worldview in terms of which we seek out and interpret new information, even when trying to formulate a more coherent belief system that could alter the previously existing system. This "hermeneutical" or interpretive process affects both rational and irrational methods of belief formation, both of which are strongly motivated.

Many recent theories about moral emotions especially emphasize the prominence of the "confabulatory" aspects of the emotions in relation to ethical, social and political issues (here again, prominent examples are Lakoff 2008; Westen 2008). They point out that we tend to believe what we wish to be true. For example, confabulatory belief systems can be motivated by the need to reduce cognitive dissonance (Festinger 1957). If the facts don't fit what we want to believe, then we tend to tweak, reinterpret, or "frame" the facts as needed. Obviously, this is an important point that needs to be incorporated into any moral psychology. Daniel Goleman (1996) presents a credible assessment of the way "vital lies" are constructed to help us navigate our way through some of the more disturbing truths of life.

But if confabulation were the whole story, it would be difficult to imagine why anyone would ever believe anything that they wish were not true. If it makes me feel good to believe that "The Voice of God told me to commit murder," then why not just believe that this is the case? Why do only psychotics believe things for this kind of reason?

It would be hopeless to try to explain the love of truth as merely an *instrumental* value whose purpose is to lead indirectly to situations that maximize rewards or minimize punishments. Suppose it were up to me to compute whether a belief that "the Voice of God told me to commit murder" is more or less likely to maximize my pleasure in the long run: Before I could make such a calculation, I would first have to assess whether certain facts are *true.* Mother Nature must have anticipated this dilemma when she designed us (or most of us) so that the exploratory drive (and thus a desire to learn the truth) is substantially independent of the other drives – not merely instrumental to their service.

But why can't a behaviorist or learning theorist just explain this situation by saying that we accept unpleasant truths in the short run because we have been conditioned to expect, *in general,* that doing so is usually likely to lead to more long term happiness or well-being?

Here too, this explanation would presuppose that there had already been some reality-based calculation of this fact. But more important, the recent research indicates that sophisticated animals, including all mammals, are motivated to explore, regardless of whether such behavior leads to some other desirable outcome, and even if the exploratory behavior is somewhat in conflict with the satisfaction of other important motivations. That this natural drive can then be *selectively* suppressed leads to more complicated scenarios.

The argument is often made that moral and political judgments are "irrational" because it is the emotional rather than the rational parts of the brain that "light up" in brain scans when people argue about those topics (for example, Shemmer 2010; Emory University Health Sciences Center 2006). But to see how flawed this inference is, consider what would happen if a subject were presented with an interlocutor who argues that the earth is actually the center of the universe, and plans on taking action that would result in loss of life on this basis (as the Pope did when ordering the execution of Giordano Bruno in 1600 for contradicting the Church's geocentric worldview). If a modern-day pope took that position, which parts of our subject's brain would light up? Would someone need to use the reasoning parts of their brain to rethink their long-held belief that the earth goes around the sun *during the real time* of this argument? And if not, does that prove that their beliefs about such things are "irrational"? Or would it prove only that they made their *rational* judgments at an earlier time, and now are reacting emotionally to someone who disagrees?

Hibbing et al. (2013) show that political liberals and conservatives exhibit different levels of "negativity bias" (perception of stimuli as threatening or harmful); the increased negativity bias of conservatives in turn correlates with a brain structure that supports a tendency toward hyperactivation of the amygdala. But we should not jump from the fact that different brain structures support different ways of thinking to the conclusion that political belief formation is entirely irrational. We know that different styles of both correct and incorrect *mathematical* thinking are supported by different brain processes, but this does not imply that mathematical thinking is always irrational.

Even the strong role of emotion in ethical inquiry doesn't mean that the inquiry is "irrational" or guided only by emotional whims or "preferences." If the only role of emotions were to cause us to believe what we wish to believe, we would face the following paradox: I wish it were the case that I don't believe whatever I wish to be true, but the psychologists tell me that I always do believe whatever I wish to be true. So therefore I must now believe that I always believe whatever I wish to be true, even though I wish I didn't have to believe this. If the motivation to confabulate were the whole story about the relationship between emotions and beliefs, then it would be self-contradictory to believe that this very thesis is true.

But as we have already seen, Hume's "love of truth" is grounded in the same emotion system that motivates epistemic curiosity in other contexts. In Hume's view, fellow feeling doesn't motivate the love of truth in the moral realm, but just the contrary: The love of truth motivates us to ask whether fellow feeling ought to be universalized, and to what patterns of behavior it ought to lead.

If the love of truth is motivated by an independent exploratory drive, then our moral beliefs may not be pre-ordained by egoistic, altruistic, or mutually cooperative feelings. We may be wealthy, yet eschew political belief systems skewed toward benefitting the wealthy. At the same time, the SEEKING component of emotion can preserve the autonomy of ethics against the flights of fancy that our everyday feelings of empathy and sympathy might engender. Those overly specific empathic feelings could result in too much partiality toward our own friends, tribe, or social class, and too little concern with the welfare of people outside of that narrow circle.

Our emotional repugnance toward the homeless psychotic, whom God may have ordered to go naked or to spit in our face, is not likely to lead to benevolent crusades on his behalf. But we can still believe that the person's suffering ought to be alleviated, and thus feel inspired to take action. The love of truth, if not dependent on other motives, can motivate us to believe things that are not necessarily in the service of our other emotions, and therefore can ground a motivation to be moral which at the same time doesn't dictate the *content* of our moral beliefs.

2.4 The Darker Side of Irrationality

In opposition to the Humean "love of truth" tendency is the equally powerful tendency to fabricate belief systems that play all too conveniently into our hermeneutical need to "interpret" reality to fit a preferred and already-established worldview. As I have stressed repeatedly, we all necessarily use a hermeneutical process in the continuing attempt to maintain overall coherence in our worldview. We all interpret new information to try to make it cohere with a previously established overall worldview. But when the balance is shifted decisively toward denying realities that challenge the presupposed position, rather than rethinking aspects of that position, it is easy to see how potentially unlimited degrees of irrationality could result. The crucial question is not whether the "love of truth" can be overcome by the tendency to confabulation. The question is how the conflict between the two motivational processes is resolved.

The issue here is not merely threats to *specific beliefs* – for instance, that my minister cannot possibly be guilty of the crimes with which he is charged. For our purposes, the larger issue is how we respond to threats to our entire worldview. Ernest Becker (1973) describes the way belief systems can be constructed to deny existentially threatening human realities. Greenberg et al. (1997) developed this approach into what is now called "terror management theory." In this way, we are motivated to confabulate authoritarian and other "conspiratorial" constructions as aspects of our entire worldview. We do so to hide from ourselves the most terrifying facts of human reality, and the tendency to do so is intensified by situations that exacerbate such terror, such as extreme social isolation or the terror of one's own death.

A classic example that fits this theory is the phenomenon that Adorno and Altemeyer describe and measure under the rubric of "authoritarianism." The authoritarian's preconceived worldview is notoriously resistant to change, despite any countervailing arguments or evidence. Also consistent with the "terror management" concept is the ease with which the authoritarian worldview can be used to reduce or avoid internal conflict and to deny anxiety-arousing realities. Authoritarianism, as measured by Adorno's Authoritarianism scale, and also Altemeyer's "Right-Wing

Authoritarianism" scale, statistically correlates with many unrealistic attitudes that simply deny facts perceived as threatening (for example, see Altemeyer 2008). Like the early Adorno with his Authoritarianism research, Altemeyer is a statistics "wonk," and the correlations that both researchers demonstrate are careful, reliable and well-controlled. Such correlations have been replicated many times, for example by Duriez and Soenens (2009), Hodson et al. (2009), and many others.

From Becker's perspective, the most troubling issues that humans must face are the threats to our power, importance, survival, and ability to fit into a social framework. All these needs are crucially reflected in the various basic emotions catalogued by Panksepp. Each aspect of the authoritarian worldview works to reduce anxiety around these often terrifying issues. For example, extreme religious fundamentalism attributes most evils to human actions rather than to the fundamental nature of reality – thus maintaining a sense of control over our destiny. Extremist political views foster the same feeling of substantial self-control, even while inconsistent with many empirical facts (for example, the problems of the poor, the need for regulation of the finance system, and sometimes even evolution and global warming).

For the same reasons, Altemeyer shows that Authoritarians are even susceptible to far-fetched theories about the inferiority of other races and nationalities, who are believed to be an economic and criminal threat, and a challenge to the purity of the national race. A belief in one's innate superiority alleviates anxieties around personal insignificance and powerlessness – issues that can raise troubling philosophical questions. Authoritarianism also statistically correlates with a comparatively violent, victim-blaming worldview (see Altemeyer 2008 as well as Adorno et al. 1964 and Hodson et al. 2009). Here again, violence, or even vicarious identification with a forceful cult leader, make one feel more in control of one's destiny. Disparagement of an outgroup also provides a scapegoat for problems, which again eases the anxiety of acknowledging the real extent of one's own relative powerlessness in the ultimate scheme of things and one's inability to solve problems or attribute them to their real causes. Correlatively, no one can compete with the "authoritarian personality," as assessed by both the Adorno and Altemeyer authoritarianism scales, when it comes to the wide distribution of ethnic, religious, or other kinds of personal animosity – a correlation again replicated by Altemeyer (2008), Duriez and Soenens (2009), and Hodson et al. (2009). Hodson et al. also use the "Big Five" personality inventory to show a huge negative correlation between Social Dominance Orientation and the "agreeableness" factor on the Big Five.

And in the classic authoritarianism research, the correlations to aggressiveness and use of force are remarkable.

Also consistently with the "terror management" hypothesis, authoritarianism includes a tendency to *polarize* the valuation of others. Since it would be terrifying to believe that misfortune might be inevitably written into the fabric of things, it must instead be the fault of someone's misuse of their free will. Thus everyone tends to become either all-good or all-evil, the latter morally deserving of mistreatment. This is especially clear in the research of Altemeyer (2008), who continues to update the empirical research into this kind of hermeneutical distortion of thinking.

Altemeyer's data might be stronger still if his measurement instruments didn't conflate authoritarianism with trust in the *government* as an authority (as reflected in his "Right-Wing Authoritarianism" questionnaire). Obviously, many authoritarians choose certain political or religious leaders as authority figures, and then when those figures are out of power, they lose their trust in the government and become its bitter opponents. The distrust in that case is precisely because the government is now the enemy of the chosen authorities. Hence the merely apparent paradox that the "Reagan revolution" in American politics, which declared "too much government" itself to be the problem, was overwhelmingly supported by authoritarians. "Get the government off the backs of the people" was one of Reagan's favorite slogans.

Hodson et al. show that the subset of authoritarians who score the lowest on Social Dominance Orientation also score surprisingly high on the "Agreeableness" dimension. This finding underscores our earlier point that authoritarians can be very nice people when they want to be. We can't just say that they generally lack empathy. The overly aggressive and simplistically judgmental attitudes toward immigrants, Jews, liberals, etc., emerge primarily when the authoritarian's social and moral beliefs are challenged.

We can therefore see that a more basic conflict underlies the authoritarian's process of belief formation, involving the hermeneutical process of interpreting reality to make it fit the pre-established rigid belief system. According to Becker and Greenberg et al. (1997), the purpose of such a belief system is precisely to protect against the terror of human frailty and basic life challenges in general, and is particularly exacerbated in the face of actual life problems, such as the threat of complete social isolation in a shifting economic structure. Such beliefs serve the purpose of "terror management" by protecting a rigid overall worldview, but it also works against the truth-seeking motive, especially in social and moral contexts.

As Gadamer (1971/1975) stresses, *we all* interpret new experiences to try to fit them into an already-existing overall understanding of reality, but the authoritarian has more difficulty with this process than others do. So it can be helpful for a general conflict approach to moral psychology to explore the ways in which psychological terror manifests itself in moral contexts.

We should not ignore the complicated and often controversial pros and cons of the "authoritarianism" concept in psychology. It is difficult statistically to separate out authoritarianism from conformism, for example. Authoritarianism also sometimes serves as a proxy for "Social Dominance Orientation" – a tendency to endorse a rigid power hierarchy in which everyone's place is pre-defined and unchanging (see Duriez and Soenens 2009). Despite these complexities, the correlations Adorno and Altemeyer have found between authoritarianism and other beliefs, attitudes and personality variables remain robust, and this must reflect certain tendencies in this kind of belief system. Adorno and Altemeyer find that authoritarianism consistently correlates with racism, extremist ideologies (especially at the extreme right-wing), religious intolerance, a tendency to solve problems through force rather than reason, and many other beliefs and attitudes.

Hannah Arendt paints an even more disturbing picture of the irrational tendency toward destructive social and moral viewpoints in *The Origins of Totalitarianism* (1968). In addition to the points raised by Adorno, Becker, and the hermeneutic philosophers such as Gadamer, as well as cognitive dissonance theorists like Festinger and conformism theorists like Stanley Milgram (1974), Arendt also sees social irrationality as linked to a certain form of implicit *value nihilism.* The utter irrationality of popular support for a totalitarian movement, she argues, does not stem from any particular political interest or policy opinion on the part of the masses, as normal political behavior does. Instead, the popular upswelling support for the radically irrational political movement, whatever it happens to be – Nazism, fascism, or Stalinism in Arendt's day – is driven by masses of people who previously were outside of the political process altogether. The only sentiment that they all share in common, in Arendt's analysis, is that frustration with the way their lives are going has reached a tipping point – in essence, the point beyond which we stop looking for solutions to problems and begin seeking excuses for our *inability* to solve them, and for some outside force on which to blame them.

The frustration in her view is not only economic, but also involves a psychological feeling of insignificance, anonymity, meaninglessness, a feeling of being "superfluous" and "expendable" in a mass society (Arendt 1968, pp. 316ff). The masses in this case are no longer driven by either altruistic or self-interested values; instead, devotion to the totalitarian movement now takes priority over everything, even including self-interest (p. 318). The mass support that is necessary for totalitarianism comes less from people seeking solutions for their problems than from those seeking to express their anger and to find a philosophical rationale for the perceived failure of their lives, including and especially their feelings of social isolation.

The role of feelings of social isolation and lack of "belonging" in Arendt's description of the rise of totalitarianism, written just in the aftermath of World War II, is strikingly parallel to Panksepp's concept of the PANIC system. When we are socially isolated, emotional pain is triggered because of agitation in this basic brain system. We are naturally hardwired to need social attachment. When people in complex industrial societies lose their sense that they are needed parts of a social community – when they are merely replaceable cogs in the machinery of production – they feel, as Arendt puts it, "superfluous" and "expendable." It is natural at that point that when a mass movement offers them a home, a community that seems to need and value their efforts, they would be highly tempted by the offer.

This solution to the problem of PANIC stemming from the feeling of social irrelevance (if Arendt's analysis is correct) could easily come into sharp inner conflict with the SEEKING system. The "love of truth" is still urging them to consider the rational coherence or incoherence of the worldview that is being offered them, but to a greater or lesser extent it must be selectively suppressed in favor of relieving the PANIC. How this conflict is resolved, and why it is resolved in one way rather than the other, is the more difficult question that we will need to get to.

What Arendt is describing is more than just irrationality. If the phenomenon that Arendt sees is real, it stems not just from a distortion of the hermeneutical process, but also and more fundamentally from an underlying implicit value nihilism. Absent value nihilism, people would generally want to at least *try* to solve problems rather than only to find excuses for not solving them; and the latter seems to be the essence of the totalitarian mindset that Arendt is describing.

Value nihilism is an important concept to be explored for moral psychology, in the same way that injury to a part of the brain can provide clues as to the *normal* functions of that same brain area, by observing which normal behaviors *fail* to occur. In fact, if the feeling of being "superfluous" and "valueless" can reach the tipping point of an implicit value nihilism, there would be even greater danger of this outcome at a time of too-rapid change, when the skills and services people have to offer become ever more quickly obsolete as they are supplanted by new technology and the effects of huge corporations with their "economies of scale" – in other words, times like our own. So these factors may be as relevant now as in Arendt's time.

The feeling of "valuelessness" is not necessarily "irrational," even though it may involve a confusion between instrumental and intrinsic value – the confused attempt to define our own value relative to some outcome toward which we are merely instrumental. Obsolete workers feel "useless" because they don't feel that they serve some important value. This dynamic complicates simple irrationality by adding to it an underlying value nihilism or semi-nihilism. For now, our central focus is the conflict between the exploratory drive and the irrational tendencies per se.

In the analyses of Becker, Adorno, Altemeyer, and Arendt, authoritarianism includes scapegoating, rigid us-versus-them polarizations, and religious or political ideologies that justify demonizing and punishing those whose interests happen to conflict with our own. The purpose of this demonization may be at least in part to deny the value of beings with whom it is cognitively easier not to empathize too much. Empathizing with the enemy creates anxieties, doubts, questions, and inner conflicts, even though we are easily *capable* of affirming the value of people with whom we are in conflict. These inner conflicts remain real, even though we also are endogenously *motivated* to consider the question whether we should universalize our compassion (as driven by the "love of truth"). In the thinking of Becker and Adorno, we tend to suppress our truth-seeking motivation in these cases in order to feel better about inflicting hardship on those whom we have managed to demonize or diminish for the sake of avoiding terrifying realities.

To use the simplest example, it becomes easier to kill enemies if we can persuade ourselves that they are morally evil, or belong to a racial, ethnic, or religious group whose members have less value per se. Similarly, it may well be that one of the most basic motivations to authoritarianism is to elaborate a worldview that justifies cutting off empathy in just those selective cases where doing so reduces cognitive dissonance and thus makes us feel better by reducing our anxiety (including not only anxieties from FEAR, but also from PANIC). How do we kill someone even while recognizing their humanity and acknowledging their intrinsic value? How do we accept the slings and arrows of fate without finding someone to blame for them? Opening such questions can be hermeneutically threatening. It can threaten the integrity of an already existing overall worldview.

There are ongoing criticisms and modifications of the "authoritarianism" concept, many of which are well managed in Altemeyer's current authoritarianism research. The point for now is that, however we theoretically conceptualize the psychology of this kind of belief system – whether in terms of cognitive dissonance theory, authoritarianism, Social Dominance Orientation, "terror management," etc. – such overall belief systems do involve in one way or another a motivation to confabulate in the interest of maintaining a certain pre-established worldview. They involve a motivated hermeneutical process – a way of interpreting new experiences to make them fit into an already existing understanding of reality.

Here again, we all have this tendency. But authoritarianism is an especially clear example of an extreme exaggeration of the hermeneutic process, and the attempt to cope with it in non-rational or even irrational ways. The interesting question is how the *conflict* between this tendency and the love of truth plays out in different circumstances – since both tendencies are always present in all of us. That question will be considered in the next chapter.

When we face severe uncertainty as to what we ought to do, our philosophical presuppositions are threatened. Correlatively, when our philosophical presuppositions are threatened – in other words, when our entire understanding of reality is in danger of crumbling – the temptation to avoid moral deliberation beckons even more strongly. If we can resolve an issue simply by telling ourselves that the beggar "would have spent the money on liquor anyway," or some other equally facile sentiment, we can avoid the difficulties caused by the inherent uncertainty of moral reasoning. Dealing with uncertainty therefore requires that we resist all the more strongly a tendency to reduce moral deliberation to whatever altruistic (and non-altruistic) emotions we already happen to feel.

In fact, in this case there always will be a temptation to gravitate toward the most *vague and malleable* moral concepts. For example, the concept of "moral desert" is tempting precisely because it is almost infinitely malleable: The criteria for determining who intrinsically "deserves" which rewards or punishments are so subjective, indeterminate, and controversial that the concept can easily be shaped and molded to fit our own hedonic interests and previously established philosophical presuppositions. As soon as the notion is accepted by both sides of a disagreement that "desert" is the standard for what should be done, no interlocutor can prove to us that our own criteria for "desert" are not the correct ones. After all, we can always point out that *their* criteria are just as vague and subjective as our own. We therefore easily escape the need to consider the issue at hand from a rational perspective. So it is not surprising that irrational thought processes like authoritarianism include a quickness to blame and to judge people as "morally deserving" of this or that punishment or reward.

This kind of problem inevitably must enter into the process of understanding the role of the kinds of emotions that affect the selective suppression of the truth-seeking motive and thus the shaping of a person's overall hermeneutical worldview. The very vagueness and malleability of the "desert" concept, both because of its arbitrary nature and also because of its capacity to undermine the truth-seeking function by playing into our confabulatory tendencies, is an example of the kind of question that must enter into the evaluation of the epistemic coherence of a moral belief system.

We have explored in this chapter some of the neuropsychological and neuro-phenomenological evidence for an independent truth-seeking motive, and its implications for the naturalistic fallacy problem. We also explored some of the conflicting emotional tendencies that eventually can build on each other to lead to "confabulatory" aspects of our overall worldview. It then becomes more clear how moral epistemology must confront the internal *conflicts* between the truth-seeking motive on the one hand (even as it must deal with moral ambiguity and epistemological uncertainty) and on the other hand the various distorting motivational influences on the hermeneutic formation and maintenance of more and less coherent value systems with their concomitant overall worldviews. The next chapter explores that conflict in a little more detail.

Appendix 2.1

Panksepp's Eight Unconditioned Emotion Systems: A More Detailed Breakdown

1. The SEEKING system (which forms the neurophysiological basis for what many call the "exploratory drive") energizes the body toward exploratory and generally non-consummatory behaviors. This brain system initiates the spreading of neurotransmitters and nerve signals that energize the body to act rather than just react. Physiologically, it emphasizes a circuit that runs through the *lateral* hypothalamus (*not* the medial hypothalamus, whose stimulation causes PLEASURE; see next), and includes the anterior cingulate (used for control of

attention), and parts of the prefrontal cortex. In terms of neurotransmitters, this system heavily relies on dopamine and glutamate.

2. The RAGE system extends from the medial amygdala through to the *medial* hypothalamus (rather than the *lateral* hypothalamus emphasized in the SEEKING system), also including some brain-stem areas. This RAGE system spreads neurotransmitters and nerve impulses back and forth between the amygdala and the periaqueductal gray (PAG) of the deep subcortex. The neurotransmitters involved emphasize substance P and acetylcholine activity.

   Importantly for our purposes, RAGE and FEAR *inhibit* the SEEKING system. Conversely, RAGE and FEAR occur especially when external events *curtail freedom of action* (and thus curtail the SEEKING system). The quickest way to elicit RAGE from an infant is to bind its movements. When the infant flies into a fit of rage, this brain system and its neurotransmitters go into action.

3. The FEAR system extends from central and lateral amygdala (again, primarily the "old amygdala") to the medial hypothalamus, to brain stem areas including PAG and pons (with signals and neurotransmitters running in both directions). The neurotransmitter activity emphasizes glutamate, monoamines, and many neuropeptides.

4. The LUST system runs from PAG through stria terminalis to corticomedial amygdala to ventral-medial hypothalamus (in females) and preoptic hypothalamus (in males), using vasopressin, oxytocin, and steroids.

   Oxytocin subserves loving feelings, as it also does in the CARE system (next). *Vasopressin*, paradoxically, is activated in male competitiveness but also in the male's gentle feelings toward the female. And the male uses oxytocin in tender feelings toward both his mate and their offspring. So we can see that there is some overlap between the LUST and CARE system (listed next).

5. The CARE (nurturance) system runs from the PAG of the deep subcortex, through stria terminalis, to ventral tegmental area (VTA), to anterior cingulate and preoptic areas. The main neurotransmitters it uses are oxytocin, prolactin, dopamine, and various opioids. Oxytocin, opioids, and prolactin seem crucial for the subtle feelings of acceptance, nurturance, and love.

6. The PANIC/SEPARATION DISTRESS (social bonding) system runs (in both directions) from the bed nucleus of stria terminalis to dorsomedial thalamus to preoptic hypothalamus to PAG, also including

in primates the anterior cingulate and anterior thalamus. In terms of neurotransmitters it relies on opioids, CRF, and acetylcholine, as well as many of the same neurotransmitters used in the LUST and CARE (nurturance) systems.

7. The PLAY system can easily be observed in young animals who instinctively play without any external reinforcement of the play behavior. This system runs from midbrain central gray to posterior thalamus, relying extensively on opioids and acetylcholine.
8. The PLEASURE/PAIN system (pleasure in the sense of consummatory pleasure, and pain in the sense of displeasure arising from bodily imbalances) runs from PAG to medial hypothalamus (not lateral hypothalamus as in the SEEKING system), and relying on opioids, as regulated by GABA. This system is not necessarily activated or involved in producing the behaviors of SEEKING and PLAY, which Panksepp characterizes as "non-consummatory" – i.e., not in need of external reinforcement with rewards and punishments.

NOTES

1 Some might think we could simply avoid the "ought implies can" problem just because, as an empirical fact, we *can* want to promote others' well-being – regardless of what *motives* are driving us to do so. Thus I *can* choose to do what is right; so doing only what I "want" doesn't prevent moral concern after all. But the problem with this approach is that it still holds that we ought to enhance others' well-being, *only when we are motivated to do so.* The problem remains that we ought to enhance others' well-being when and if we want to. I.e., we always ought to do whatever we want.

2 In earlier discussions (for example, Ellis 2005), following Panksepp's earlier terminology, I combined the LUST and CARE systems into one "nurturance/sexuality" system, because Panksepp (1998) emphasizes the strong overlap between these two brain systems (as can be seen from the above roughly simplified breakdown of the physiology involved).

3 In fact, these other motivational circuits generally *inhibit* play. While a certain amount of activation of dopamine SEEKING circuits seems necessary for play, the main reason for this is that, in a new environment, animals conduct exploration before they begin to play.

4 The extreme example of this lack of self-transparency is "alexithymia" – an inability to read the meaning of one's own feelings. One result of the "experiencing scale" (Klein et al. 1969), as well as other more recent measures of alexithymia, is that people in effect suffer from various *degrees* of alexithymia in various situations – alexithymia is not an all or nothing (for example, see Gendlin 1978/1982).

CHAPTER 3

Love of Truth and "Vital Lies"

Basic Conflicting Emotions in Moral and Political Psychology

The difficult question is not how we can fall into moral and social irrationality. The real question is how the *balance* between this tendency and an equally basic "love of truth" is resolved.

Notoriously, large numbers of Europeans who suffered from the bubonic plague during the Middle Ages blamed the plague on a Jewish conspiracy, despite the absurdity of such a far-fetched "conspiracy theory," and as a result thousands of Jews were burned at the stake. Such demonizing of an "outgroup" is doubly typical of the Adorno/Altemeyer "authoritarian personality" dynamic. Not only does it express ingrained prejudices against those outside the authoritarian's preferred social framework; it also allows dogmatic beliefs to remain unquestioned.[1] In addition to the need to vilify an outgroup, the persecution of the Jews also expressed the tendency toward violence and solving of problems through sheer force that Adorno's and Altemeyer's data suggest is favored by authoritarians, who themselves may have been indoctrinated by means of the threat of force and/or humiliation as children – an etiological connection suggested by George Lakoff (1996), Yehuda et al. (2001) and others.

Yet sometimes people do *not* blame plagues on a Jewish or other such absurd "conspiracy." Sometimes they turn to scientific medical research, with reasonable understanding of its moderate levels of uncertainty. Other times, they at least admit to themselves: bad things can happen to good people. So we need to ask ourselves what determines this difference. Why does the "love of truth" sometimes win out in our internal tugs-of-war, and other times confabulation takes the day?

For our purposes, it isn't necessary to argue about whether authoritarianism is a personality "trait." It may in fact be a tendency that we all exhibit to varying degrees in varying circumstances. It often may also serve partially as a proxy for simple conformism, or "Social Dominance Orientation," or the Becker/Greenberg "terror management" dynamic that we discussed earlier. Or the same social attitudes may simply result from cognitive dissonance. It is sometimes argued that up to 60 percent of Milgram's subjects may have delivered the painful electrical shocks, *not* because of either authoritarianism or conformism, but rather because of "belief perseverance"; they could not easily *change* their perception of the benevolent experimenter to an image of a malevolent one (Moti 1990). The important point for our purposes is that, from the standpoint of a conflict approach to moral psychology, we are dealing with one or another form of confabulatory tendency in social and moral belief formation. And this tendency is in conflict with the equally natural truth-seeking motive.

From the terror management perspective, Becker, Solomon and Greenberg stress that the Europeans who blamed the plague on the Jews couldn't admit to themselves the horrible truth of their helplessness. To the extent that "terror management" is the issue, Solomon and Greenberg suggest that the Christian Europeans' previous worldview couldn't allow that the plague was an indication of humanity's horrifying weakness and vulnerability, of our infinitesimal smallness and powerlessness. Their terror in the face of such troubling truths was too much for them to face. Doing so precipitated too much cognitive dissonance and was too much of a challenge to their habitual worldview. In such a worldview, it seems inconceivable – and emotionally intolerable – that God would have allowed such a horrible pestilence unless someone had misused their free will and maliciously conspired to create it.

But despite such examples, which history brings all too easily to mind, if the motivation to feel good, to ease our anxieties or to reduce cognitive dissonance were the whole story, then we would always simply believe whatever made us feel better regardless of any evidence. What we are dealing with is a basic inner conflict between different motivations that are equally primordial.

Even from the standpoint of Mother Nature engineering an organism with robust capacity for survival through intelligence, it would make sense to protect the truth-seeking motivation by making it as independent of the feel-good motivations as possible. Hume's idea of a "love of truth" can now have the extra credibility of being describable in neuropsychological terms as well. It is part of Panksepp's SEEKING system.

Neuropsychology is now beginning to fully appreciate the extent to which this system is independent. Panksepp puts it this way: "The extended lateral hypothalamic (LH) corridor [with ascending dopamine circuits and descending glutamatergic circuits] . . . responds unconditionally [i.e., without any previous learning or conditioning] to homeostatic imbalances"

(Panksepp 1998, p. 145). As a result, "this harmoniously operating neuroemotional system drives and energizes many mental complexities that humans experience as persistent feelings of interest, curiosity, sensation seeking, and, in the presence of a sufficiently complex cortex, the search for higher meaning" (p. 145).

Panksepp's characterization of this motivation as a *non-hedonistic* one, and his consequent rejection of purely hedonistic reinforcement theories, is quite explicit: "This is a diametric reversal of traditional behaviorist thinking.... The affective state [that serves as a motivator in some instances]... is not simply 'pleasure' but a highly energized state of psychic power and engagement with the world" (Panksepp 1998, pp. 147–150).

Some learning theorists might suggest that exercising the exploratory instinct *is itself* a form of "pleasure." Even Frijda (2006/2007) sometimes speaks in these terms (ch. 4 and passim). But the notion of "pleasure" as a reward for behaviors that *would not occur unless rewarded* is very different from a concept of "pleasure" in which the behavior is *its own* reward, and would still occur even if not instrumentally rewarded or even if punished. Nor are such behaviors necessarily experienced as pleasant, as in the case of birds instinctively migrating hundreds of miles without rest.

From a phenomenological point of view, there is a simple way to notice in our everyday experience that the love of truth may motivate us to believe statements based on their truth value, regardless of whether believing in those ways makes us subjectively *feel* better (in either the short or long run). Sometimes we reluctantly acknowledge that, previously, we *were* believing a false or confused statement or interpretation of reality simply because it made us feel better to believe that way – and we then find ourselves wincing in our feeling of disturbance over such "inconvenient truths." The emotions that motivate us to admit to ourselves that we were deceiving ourselves must be different from the emotions that motivated us to deceive ourselves in the first place. Nor is it a pleasant feeling to be forced to change a previous opinion in this way.

The exploratory drive can function independently of pleasure maximization in much the same way that salmon swim upstream to their original spawning areas, without any reward to the individual organism for doing so, other than the behavior itself (which actually is quite grueling and unpleasant, not to mention the further unpleasantry that it ends in death; as for the opportunity for sex, the individual fish doesn't know that this consolation awaits at the end of his grueling journey). The behavior is a hardwired instinctual tendency that doesn't depend on reward.

3.1 Exploratory Drive, "Love of Truth," and the Roots of Moral Conflict Psychology

As an undergraduate student, I was confused by the notion of a conflict between the "Reality Principle" and the "Pleasure Principle," because the theories to which I was exposed at the time seemed to imply that the "Reality Principle" is always merely a skill set *in the service of* the Pleasure Principle or other non-rational emotional purposes, if sometimes only indirectly. How can one principle that is merely in the service of another one also be in conflict against that other one?

But what the recent emotional brain research confirms is that the Reality Principle is not just a skill set. It is an independent *motivational* system. From an evolutionary standpoint, the main reason we can sometimes behave in certain ways independently of whether they maximize our pleasure or "reinforcements" is because natural selection led to the survival and flourishing of *whole species* in which individual organisms engage in exploratory behavior irrespective of whether doing so brings rewards to the individual organism.

Peter Singer's (1981) analysis of the sociobiology of human behavior is partially in agreement with Panksepp on this point. But the new research now shows us the specific neurophysiological basis for how this works in the motivation of the *individual organism*. It is not simply that (as Singer implies) we are hedonistically motivated to use impartial moral arguments to *persuade others* of the truth of ethical propositions, presumably because such persuasion benefits ourselves. Rather, we want to know what the truth *really is*, just for the sake of knowing. And we need to know this, in our own minds, especially in those cases where we have no practical necessity to, as Singer puts it, "justify . . . conduct to society" (Singer 1981/2011, p. 100).

We see this point played out especially well in the Camus novel *The Fall* (1963). No one other than the protagonist himself knows that he once refrained from saving a drowning person – there were no witnesses to the event, and he has no need to justify his inaction to anyone but himself. He tries to pretend even to himself that he feels no guilt. Yet he is so unable to live with himself that his life from that point forward is a downward spiral.

The crucial difference between consummatory and non-consummatory tendencies can be seen neurophysiologically: Consummatory pleasure relies more on the use of morphine-related brain chemicals, whereas seeking and play rely more on dopamine, norepinephrine and acetylcholine activity. And stimulation of different parts of the hypothalamus (which used to be thought of as the "pleasure center" of the brain) leads to activation of two *entirely different* brain systems: The PLEASURE system (as in the Olds "pleasure center" studies – see Olds 1956) is activated if we stimulate one hypothalamic area, whereas the SEEKING system (as in many studies carried out by Panksepp and others) is activated when a different part of the hypothalamus is stimulated. Many of these studies are well summarized in Panksepp (1998, ch. 6).[2]

Panksepp's main objection to behaviorist learning theories is that, on the one hand, they need to distinguish between consummatory and instrumental behavior, in order for the reinforcement to "explain" the behavior. But in non-consummatory emotion, the behavior is essentially its own reward and doesn't need to be extrinsically reinforced, nor does it even result from a "habit" indirectly derivative from extrinsic reinforcement (as Hull 1943/1966 would have had it). Not all behaviors (for example, the search for truth) need to be reinforced by anything other than the behavior itself.

In defense of behaviorism, it can be argued that play, nurturance and exploration could simply be regarded as primary reinforcers – and thus incorporated into the behaviorist paradigm. But then, as Robert White (1959) pointed out so clearly many years ago, the distinction between behavior and reinforcement becomes blurred, and explanations of behavior in terms of reinforcement become non-explanatory.

The relevant point is that there is clearly a difference between behaviors that we engage in for their own sake, and those that only lead to an end, such as maximizing our pleasure. This was reflected in empirical findings as long ago as Harlow (1950), Kagan and Berkun (1954), Montgomery (1954) and many others reviewed by White (1959). In each of those studies, a non-consummatory behavior was its own reward – a monkey rewarded by being allowed to look out the window; chickens that literally would rather peck than eat; and rats that do work in exchange for the reward of being allowed to press a bar. During the twentieth century, these odd cases were treated, in Kuhnian terminology, as "anomalies" as yet to be explained rather than as indicating the need for a new kind of theory.

Even from a behaviorist point of view, there must be a distinction between behaviors that need extrinsic reinforcement and behaviors that serve as their own reinforcement. What is clear is that we can't explain endogenous behaviors like exploration or nurturance as merely "secondary" reinforcers, derivative through learning from extrinsic rewards. In short, some of our basic energy systems inherently seek to explore and to be active, curious, and playful even when these tendencies are not reducible to a more basic Pleasure Principle (see also Ellis 2005).

3.2 Easy versus Hard Problems of Internal Conflict

If the "pleasure principle" and the "love of truth" are independent drives or tendencies, we shouldn't conflate the *easy* question about this conflict with the *hard* question. The hard question about confabulatory belief systems like authoritarianism and the blaming of the plague on the Jews is not merely to understand how we could be motivated to confabulate. The more interesting question is: What would determine whether the "love of truth" is subordinated to the pleasure principle (as served by confabulation) or vice versa in the first place? We are now considering a *conflict* between two equally natural motivations.

While the socially oriented emotion systems do motivate empathy with *certain specific* family or group members, we have already seen how inadequate they would be, if not for the SEEKING system, to ground anything like a non-trivial moral psychology. By contrast, the exploratory drive, when combined with the extensive prefrontal capacity (or truth-seeking *skill*) of humans, can lead to serious moral deliberation. If this deliberation results in the universalization of the valuation of other conscious beings, then it also mandates the project of balancing questions of beneficence, distributive fairness, and other moral principles. As promised earlier, we will have to clarify the range of meanings of terms like "ought" and "value" in our next chapter, and accommodate the diversity of these meanings.[3]

But first, let's formulate the internal conflict in more tangible terms: The *extent* to which the love of truth overrides other motives can be a variable rather than a constant. It may depend partly on the ways in which we have previously trained and conditioned ourselves, based on our own knowledge of our own psychology. Within certain limits, we can condition ourselves to be more moral than we initially found ourselves inclined to be.[4]

To even entertain the possibility that some of these universalized ethical propositions might be true could threaten to complicate interpersonal, political, economic, or even military conflicts of interest with other people and groups. This situation can lead to *internal* conflicts reflected in cognitive dissonance and related hermeneutical effects on our thinking. These effects in turn can lead us to construct belief systems to justify either the demonization or simply the ignoring of others – for example, when we need to avoid the pain of vicariously regretting their unavoidable suffering, or the pain of having to actually harm them in war or in economic competition; in many cases we construct such belief systems in order to, in effect, reduce our cognitive dissonance.

To be sure, a mere *motivation* to be moral does not yet move us to an ability to *find out* what is moral. But the exploratory drive is what *motivates* us to pursue such a possibility, which couldn't be motivated by other natural emotions such as nurturance and fellow feeling.

So now we come to a more difficult question – the truly "hard problem" of internal conflict. What determines which side of this conflict wins out in which situations?

We can take it as a well-established empirical fact that confabulatory belief systems of this kind are *not* merely the result of ignorance or lack of intelligence. We have already seen that racism and Nazism, along with the convoluted logic and empirical theories needed to support such ideas, were shared by many of the most intelligent and educated people of their times. In retrospect, we now think of such beliefs as analogous to "conspiracy theories" because of the convolutions of logic and absurd ad hoc hypotheses needed to prop them up. Arendt (1968), similarly to Luce's (1960) study of the White Citizens Councils, stresses that the appeal of irrational theories of racial superiority and other far-fetched "conspiracy theories" had just as much appeal to intellectuals and other respected professionals during the rise of Nazism as it did for "the masses."

Nor were these intelligent people lacking in empathy and compassion. As we saw earlier, the same individuals devoted themselves to charity work and could be quite generous in helping their friends and neighbors, and even the indigent. What was missing in the motivations that shaped their worldview was not empathy or compassion, but rather a willingness to ask questions and seek truths that potentially threatened an already established worldview, challenges to which created fear and anxieties of the kind Becker explores.

But we can't just explain the motivation for incoherent moral viewpoints by saying that people have a need to direct their attention away from the problems of others, for whatever reason. We all need to maintain some selective inattention to all the vast suffering going on around us, simply in order to preserve our own sanity. A physician making an incision or a soldier forced to kill during war may need to cut off empathy. But this doesn't mean that we forget that there are literally billions of people and animals in distressing situations, empathy with whom *would* be quite concerning if we *were* to focus attention on it. In short, we understand the concept that their value might be a real dimension of their being, to which attention could be directed at any time, just as the inside of a refrigerator is still there even when we close the door. We may therefore act in accord with such "affordances of empathy," even if we don't specifically feel the empathy in each individual case.

By contrast, there is often a temptation to cut off empathy by means of self-deceptive and casuistic "conspiracy theories" – attempts to defend incoherent systems of morality. This leads us directly into the internal conflict that shapes coherence and incoherence. Incoherent and "inauthentic" moral thought processes often are designed to make us feel better about this internal conflict. By using the term "inauthentic," there is no need to take on all the specific baggage of Heidegger or other existentialists. We are referring here only to emotional untruthfulness in the sense used by Salmela (2014) – untruthfulness in the interpretation of our own feelings.

We must also remember that the hermeneutical process is not an aberration. It is an inescapable part of everyone's process of belief formation. Yet as we have seen, we don't always manage the process by blatantly misinterpreting reality, even including moral reality. So the crucial question is why we *sometimes* succumb to the hermeneutical confabulations that ease these anxieties through false denial and repression, yet other times do not. The reason for the confabulations can't be a lack of intelligence or education, nor an overall shortage of the emotions of empathy and compassion.

On the contrary, what led the Nazis and the segregationists of South Africa and the United States to their elaborately confabulatory worldviews is more likely to have been a selective suppression of the exploratory drive and thus of all curiosity with respect to any question that touched on threats to an already-in-place overall worldview, and whose questioning would have rendered insecure an entire philosophical outlook. Above all else, and despite the intelligence of the person holding such a worldview, curiosity and exploration had to be held in check whenever they challenged established dogma. When the exploratory drive is even momentarily and selectively weaker than existential fears and anxieties, critical thinking comes more slowly and with more difficulty. We must also remember that even those who are merely *"conforming"* to a popular yet incoherent belief system, while they were not responsible themselves for developing the belief system, nonetheless are still succumbing to a suppression of the exploratory drive in order to avoid asking the obvious questions about the incoherent system.

So the real question – the *"hard problem"* of inner conflict – is not merely how our fears and anxieties motivate us to confabulate. We always have those fears and anxieties, but we don't always confabulate. The harder question is *why*, *when*, and *under what circumstances* we allow our fears and anxieties, and in general our cognitive dissonances, to shut down or partially shut down our exploratory system. The love of truth, after all, is sometimes stronger than the motivation to reduce cognitive dissonance – as the practical common sense of first responders in catastrophes often illustrates, even in the face of fear and danger.

We know that a part of the "old-amygdala" is activated through fear or anger, which in turn tends to inhibit the anterior cingulate (crucial for voluntary direction of attention) and to partially shut down the adjacent prefrontal activity, which then temporarily cuts down on our critical thinking capacities. Many of the brain studies on this phenomenon were done by the neuropsychologist George Bush and his associates (see Bush et al. 2000) – no relation to the US president elected in the same year. The Bush findings are corroborated by Hibbing et al. (2013), who use social-psychological research paradigms (primarily measuring "negativity bias" and the intensity of fear reactions). Hibbing et al. conclude that extreme political conservatism – which of course correlates with authoritarianism – results from a tendency to hyperactivate the fear centers of the brain, especially the amygdala. As in the Bush and Hibbing studies, the person with an extreme "negativity bias" dampens the connections between the neo-amygdala, the anterior cingulate, and the prefrontal cortex.

At the same time, there can also be a reverse effect. The prefrontal lobe can "modulate" the tendency of the amygdala to hyperactivate, and thus regulate fear or anger. (For example, see Barbas and Zikopoulos 2006; Faw 2000b, 2003; Kim and Hamann 2007; Lungu et al. 2007; O'Doherty and Dolan 2006; Gee et al. 2013.) But then what determines whether and to what extent the "old amygdala" is allowed to dominate the prefrontal cortex, or vice versa?

We can also formulate the same question phenomenologically. Almost unanimously, first responders and combat veterans who have had to "overcome" their fear in order to function rationally say that, simultaneous with the fear, there is always a more *important goal* at stake than the fear itself; and there is an available *course of action* that can lead to the goal. It is only for the sake of this more important goal that the fear, rather than being eliminated or denied, is acknowledged, admitted, and accepted as something that the emergency responder must act despite it. One allows that the feared outcome would be terrible, but we also desire to achieve the goal, at least as strongly as we fear the terrible outcome. We then focus more of our attention on the action to be taken rather than the feeling of fear, even though it is still there. Our focus shifts in large part to the environmental conditions that legitimately provoke the fear, which in most instances are the same environmental conditions against which we have to take action in order to achieve the goal. We then redirect attention to actions that can be taken.[5]

The psychiatrist Judith Herman (1992/1997) emphasizes that threats tend to become psychologically unmanageable when action is impossible: "Changes in arousal, attention, perception, and emotion [in response to threats] are normal. They mobilize the threatened person for strenuous action . . . Traumatic reactions occur when *action is of no avail* . . . " As a result, "traumatic events may sever these normally integrated functions from one another. The traumatized person may [subsequently] experience intense emotion . . . without knowing why" (p. 34, italics added).

We tend to be able to overcome fear and anxiety when we can discover a way to take action, provided that the action is toward a goal that we strongly value. But if the trauma is initially too overwhelming, and no available action is possible, the critical thinking functions are disconnected from the emotion; the action potential of the motivation in play becomes obscured from thought. At that point, we become unable to think rationally about how to confront the threatening situation. We fail to think of the actions to be taken, and why they are valuable.

Many combat veterans say that the main motivation that outweighed their fear was the need to protect their comrades. The strength with which they valued the well-being of others was strong enough to mobilize action, even though the fear was still present, urging them not to take the action. The internal conflict was resolved in favor of valuing the lives of their friends. Besides the issue of whether available actions are possible, there is also the need for a countervailing value to be felt strongly enough to counterbalance the fear or anxiety that is also simultaneously in play.

Rollo May in *The Meaning of Anxiety* (1950/2011) describes the process as one in which a driving motivation to achieve some valued objective is strong enough to outweigh the ever-present anxiety, which itself is never eliminated, and which would overcome us if not counteracted by a reason to struggle through the anxiety. In many combat situations, there is no reasonable course of action that seems likely to lead to the goal, and therefore attention to the action *cannot* overpower the terror in the face of obvious negative outcomes. Or, as Herman observes, the person attempts to act even though no action is of any avail, and perhaps survives only to suffer later from PTSD (post-traumatic-stress disorder).

Stephen Crane in *The Red Badge of Courage* vividly describes a process similar to Herman's description, in a soldier's internal fight to mobilize himself while in the throes of a terror which he never actually ceases to feel. Crane describes the inner ruminations through which the soldier eventually focuses his attention on the values in his life that will be lost if he fails to survive. He begins firing and reloading, and these actions redirect his attention. Similarly, the jazz critic Nat Hentoff (1973), in writing about the lives of jazz musicians, notes that serious musicians have a better record of overcoming drug dependencies than the general population. He ascribes this difference to the fact that the musicians in question had something to live for – their music and their families – other than merely the pleasure of a drug high. These overriding values motivated them to endure their anxieties and resist the temptation to self-medicate. Again, they were able to direct their attention toward a motivated action and away from the anxiety for which they were tempted to self-medicate.

In all these accounts, the issue is not whether fear or anxiety can be eliminated, but whether the anxiety that exists can be overpowered by a more important motivation, and whether there is some action we can take that seems like a reasonable path to the goal. We can then redirect our attention to executing the action even while in the throes of fear or anxiety, provided that some particular action seems promising. The crucial question is how we manage to balance the conflicting emotions, as opposed to eliminating one of them.

So one of the important determinants in the conflict between anxiety and the ability to take rational action revolves around the intensity with which we feel the value of something that we value, and therefore can feel the value of an action that we can take. This intensity is a function of the emotion systems that undergird *inspiration* – not the ones that undergird happiness or pleasure in the sense of instrumental reward. Moral commitment, in turn, is a function of valuing; so the strength of our moral commitment depends on the robustness with which the inspirational emotion systems, especially the SEEKING system, are activated. It is possible to value another's well-being strongly enough to inspire action even at the risk of other values. Values that are felt intensely are capable of inspiring action, essentially because valuing is a feature of acting beings.

But this process can become circular. The constant suppression of the SEEKING system, even though periodic and selective, could lead to a general damping down of inspiration in general. This could foreshadow clinical depression in some cases, but in other cases it could exaggerate moral confabulations. We should also remember that dampening of the exploratory drive can often be *selective and momentary*, not generalized. And exploration itself is only one of a variety of behavioral manifestations of the SEEKING system. The same person can be courageous in one situation, yet can also be overcome by fear of existential threats to an overall worldview in other situations. Suppression of the exploratory drive is at least as likely to be selective as it is universal.

As a result, fear or anxiety can temporarily shut down certain aspects of the SEEKING system and especially the exploratory drive. We see clear evidence of this process in the Hibbing et al. (2013) studies, where a hyperactivated amygdala response leads to an unrealistically exaggerated "negativity bias." As Bush et al. (2000) show, the same amygdala hyperactivity inhibits the prefrontal cortex, which is needed for rational functioning. If fears are so cataclysmic and threatening to our entire existence (for example, the fear of death, as Becker emphasizes) or to the entire meaning of our existence (for example, our cosmic smallness, insignificance, and powerlessness, or a complete inability to achieve satisfying social relationships, which can lead to hyperactivation of Panksepp's PANIC system), then we may tend to be overwhelmed by the fear or panic, which can lead to Goleman's (1995/2005) "amygdala overload" (in which he is actually referring to the "old-amygdala") – and the resulting "prefrontal shutdown." [6] This effect obviously could also be exacerbated by a socioeconomic situation in which no possible action seems feasible to address cataclysmic threats to people's feeling that they are valued members of the society (triggering the PANIC system), and that their lives have meaning – a complicated question that we will have to consider further. But the same kind of dynamic can occur in completely non-political and idiosyncratic contexts in which a person cannot feel the intrinsic values to be achieved by taking constructive action to solve problems.

This kind of situation inhibits the otherwise autonomous love of truth, which then allows the cognitive dissonance tendency toward confabulation to *win out over* the love of truth. In this case, we are tempted to construct ideological and religious systems to ease our internal conflicts by cutting off empathy with various categories of people, whenever empathizing with them creates too much cognitive dissonance, especially in light of Becker's "terror management" hypothesis. On Becker's view, knowledge of our own death and relative unimportance and powerlessness is so cataclysmic for all of us that it tempts us to fabricate elaborate philosophical schemes in which others are to blame for our own misfortunes; it would be too terrifying to simply admit to ourselves that such misfortunes are built into the basic order of things.

But we have also seen that a full experience of the intrinsic value of valuing creatures can inspire us enough to counterbalance these existential fears and anxieties, if we focus attention on the correlative actions we can take. Physiologically speaking, the SEEKING system with its prefrontal activities can regulate the amygdala hyperactivation by these existential fears and anxieties, and can therefore damp down our confabulatory tendencies. So here again, the real question is: What makes the difference between these two scenarios?

In the case of the selective shutdown in *moral* thinking and development, the authoritarian system offers the reward of prioritizing the Pleasure Principle over the Reality Principle by allowing the confabulation through selective shutting down of the exploratory drive. So it could well be that a need to cut off empathy per se (in selective contexts) could partly motivate the prevalence of the confabulatory tendency over the love of truth in the first place. For example, if we develop a belief system in which the enemy is evil and deserving of harsh punishment, we can kill him with less of the emotional pain that otherwise would result from our natural feeling of empathy. But then we pay the price of reducing our own appreciation of the intrinsic value of other valuing creatures simply by virtue of their being valuing creatures. This in turn has an effect on the entire inspiration function of the SEEKING system.

For now, the relevant point is that the need to cut off empathy can be a *cause* as well as an *effect* of confabulatory moral reasoning such as is found in authoritarianism. What determines which direction this circular causation moves in? There could be many contingent experiential factors – the kind of hopeless combat situations described by Herman, or even abusive childhood experiences. Many psychologists believe that PTSD is further exacerbated by having experienced such child abuse earlier in life (for example Briere et al. 1987; Herman 1992/1997; Yehuda et al. 2001). Lakoff (1996) correlates very strict child-rearing, sometimes including child abuse, to the tendency toward authoritarianism, citing Yehuda (1995) as empirical support. Such a hypothesis would also be consistent with the studies of authoritarianism and its intergenerational transmission by Duriez and Soenens (2009) and by Hodson et al. (2009). All of these findings would be consistent with the notion that an initial need to cut off empathy could in turn motivate a *selectively* non-empathic worldview, and that authoritarianism, which has the effect of cutting off empathy *only in certain sociopolitical contexts*, is one such system.

Hodson et al. (2009) further show that lack of moral concern, as reflected for example in racism and ethnocentricity, correlates with "Dark

Personality" – a combination of narcissism, "Machiavellianism," and psychopathy (the latter considered as a *variable*, not the full psychiatric syndrome). The narcissism component of this "Dark Triad" may contribute to the question of internal conflict in ways that we haven't yet considered. Narcissism features a preoccupation with one's own importance and power to affect outcomes, which also implies an especially uneasy relationship to the idea of one's own death and relative insignificance in the overall scheme of things – which is actually a universal fact of human existence. These are the same kinds of issues that figure into "terror management theory." Narcissism requires a rejection of one's own relative unimportance, powerlessness, and impermanence in the ultimate scheme. The more such concerns dominate, the more unacceptable the related threats become, precipitating a preference for denial mechanisms, which in turn can distort thinking about moral issues. But notice that we all are troubled, in different ways, by these same basic existential problems. The narcissistic concerns just listed need not rise to the level of a narcissistic personality disorder in order to affect the internal conflicts we are considering here.

For all these reasons, the motivation to cut off empathy is by no means due to an inability to feel empathy per se. The confabulatory tendencies are triggered by a need to cut off empathy only *in the specific ways* needed to preserve a person's already-established worldview. Thus authoritarianism and other confabulatory or generally incoherent moral systems can also be triggered by any situation that motivates us to cut off empathy – warfare, hard economic times, or abusive childhoods.

The Adorno/Altemeyer authoritarianism concept and the Becker/Greenberg terror management psychology offer coherent explanations as to how an entire belief system – the authoritarian belief system, for example – can be motivated by fear of earthshaking threats such as death and our own relative insignificance and powerlessness. The pre-existing worldview in the hermeneutical process has already been established in such a way as to protect challenges to these concerns. Thus a threat to that worldview is also a threat to all the other narcissistic concerns, and therefore is perceived as an existential threat of the highest importance. Any threat of powerlessness and relative insignificance in the scheme of things could also exacerbate narcissistic issues even further, which would be consistent not only with the Hodson et al. "Dark Triad" research, but also with Bell's proposed connection between narcissism, authoritarianism, and racism, as well as Arendt's analysis of the rise of totalitarian movements.[7]

For instance, as Becker describes, we may subscribe to a religious fundamentalist belief system that reassures us of easy immortality as well as an inflated assessment of our own importance on the cosmic scale, making us feel more in control of our destiny than we actually are. But then as a result we become involved in the construction of *ad hoc* hypotheses, for example the need to blame natural evils directly or indirectly on human free will. And the over-emphasis on free will then tempts us to believe either that people suffer mainly due to their own moral faults, or that some evil group such as the "international Jewish bankers" are to blame for the problems. On Becker's hypothesis, it would be too terrifying to admit that the problems could simply be a natural part of the realities that we face.

Even while such belief systems are constructed, the need for epistemic coherence is still being demanded by the basic exploratory drive. So we try to make such belief systems hang together in a more or less internally consistent way, and they can be understood in a way that is generally congruent with the fairly robust empirical correlations that Adorno and Altemeyer find between religious fundamentalism, political extremism (usually on the political right wing, which fits most neatly with the over-emphasis on individual responsibility), blaming the victim (while blaming one's own problems on "outsider" groups), and a preference for force and violence as opposed to solving problems through reason or cooperation.

We have seen so far that, if there is an ability to create confabulatory belief systems such as authoritarianism to reduce cognitive dissonance, then there must also be a countervailing and conflicting tendency. Otherwise we would always believe whatever makes us feel good regardless of any epistemic considerations. Since we don't do that, there must be something like a "love of truth" at a basic motivational level not derivative through reinforcement of any other motive.

But even if there is an independent motivation to find the truth, we still haven't established that there *is* any truth beyond the "emotional preferences" of emotivism – that a "moral realist" perspective is more coherent than a purely emotivist one, in which "values" would be only expressions of emotion, and thus essentially arbitrary social conventions or habitual behaviors indoctrinated for the purposes of the survival of the species or the "selfish gene." This question is crucial, because if moral "beliefs" are not really capable of being true or false, then the love of truth can't be much inspired by them.

Before going further, we therefore need to clarify a crucial concept that figures into this difference between emotivist and realist conceptions of value. How does the statement "X has value" relate to the idea that I or other subjective creatures "value X"? Is "X has value" a mere emotion, an assertion of fact, or some other kind of concept? The next chapter begins with a clarification of these interrelated but not identical concepts.

NOTES

1 Hannah Arendt (1966) emphasizes that the persecution of Jews wasn't merely an afterthought or a rabble-rousing technique, but an essential component of the Nazi philosophy. Jews were singled out for resentment because, given their special status as a "separate state within the state," some Jews economically prospered. This material success repelled the otherwise facile charge of "inferiority," and challenged the self-esteem of the authoritarian group; this elicited not merely contempt, but rather *resentment*, and an especially intense hatred.

2 Robert Harmon et al. (1976) describe the human infant as being born into a "distress – contentment" emotional stage based on the stimulus-averse "Nirvana principle." Then at about 4–5 months, there is a concurrent "surge in the active, curious, exploratory tendency of the infant," in "stimulus-seeking activity" – presumably the emergence of the SEEKING system – radically different from the perinatal "nirvana responses" (p 89–90; see also Faw 2015). This "stimulus-seeking activity" is set into continuing conflict with "stimulus-averse" tendencies.

3 Hume's real view is not that morality is grounded in fellow feeling, but rather that moral beliefs are based on both fellow feeling *and* the universalization of certain propositions because of our "love of truth." We extend fellow feeling, in principle toward *all* valuing creatures. "In principle," we must say, because we can't literally feel an emotion of empathy toward all valuing creatures at once.

4 We can compare this process of universalization of empathic concern to the "stages of moral development" as postulated by Carol Gilligan (1982), and which led to the "ethics of care" developed by Drucilla Cornell (1995). This similarity is ironic in that Gilligan and Cornell are often mistakenly caricatured as wanting moral sentiments to be non-rational.

5 Attention to actions we can take enables us to "modulate" the fear response and function rationally despite it. This is reminiscent of the way Gendlin describes the "focusing" process, where we take a step back and view the feeling-quality as the "felt sense" of a real situation, even though the situation may be cognitively confusing, rather than "drowning in" the internal quality of the feeling. We can then focus on environmental components of the meaning of the feeling, relative to our need to act.

6 In the case of PANIC, it is not the amygdala that hyperactivates; correlatively, grief over a separation or extreme social isolation has a very different feeling quality from simple fear or anxiety. The mutual effects of this PANIC response and prefrontal activity on each other haven't been studied as extensively as the mutual inhibitions of the amygdala and prefrontal activity.

7 In Arendt's analysis, the mass-production economic systems with their mechanistic production and monolithic investment banks lead to feelings of smallness and irrelevance for increasingly isolated individuals within the system. This feeling of personal insignificance and "superfluousness" obviously aggravates narcissistic issues. Not surprisingly, Heinz Kohut (1985), whose main focus is narcissism, therefore connects the rise of Nazism to a group narcissistic disturbance.

CHAPTER 4

Moral Realism, Hermeneutics, and Enactive Epistemology

The Truth "Resists Us"

The possibility of confabulation implies that there is also such a thing as truth. But understanding the difference between emotivist and realist conceptions of moral value requires that we differentiate between "subjective" definitions of value – in which, for example, "X has value" is equated with "I value X" – and "non-subjective" definitions, such as "it would be incorrect (in some sense) for anyone to deny that X has value." In everyday life as well as academic philosophy and psychology, people mean different things by these concepts, often depending on context. This brings us to a seemingly simple, but actually very tricky philosophical problem. How do we define what we mean by "value"?

4.1 Let's Not Play Tricks with Moral Definitions

It might be tempting to say that the "true" definition of a term is just a stipulation as to what "we" "normally" mean by the term. But history has taught that the "we" assumed here is always in conflict with some other "we," and there will always be the objection, "Yes, but others define that term in this other way." Also, the "we" in question may already suffer from the various kinds of ambiguities, inner conflicts, and distortions that we have already been discussing here.

Logically speaking, anyone can define any term any way they want to, provided that the definition is unambiguous. If someone says "I define God to mean 'the first cause,' and now I will prove the existence of God in that sense," the proof may be technically valid. But the problem is that we can't legislate that others not talk about some other meaning of the term "God." The truth is that every word has a number of different definitions in different contexts and as used by different users. The dictionary will list many different definitions. It would be a waste of time to try to argue that one definition is "the true" meaning of a term. But it would also be counterproductive to say "this is my definition, and others can define it however they like." In that case, we lose communication between people with differing philosophical worldviews.

What we can say, however, is that all morally relevant definitions of the term "value" have something in common. They all connect it to a subject valuing something, in one way or another. After all, if there were no living, conscious, sentient or in some way "subjective" beings in the universe (granting that some might count ecosystems as "subjective") – creatures capable of caring what happens – then it would make no difference whether a flood or earthquake or any other event were to occur or not. There would be no one to care. Within this broad array of definitions, we can divide the different senses of "value" into both "subjective" and "objective" senses.

When we make subjective statements, we say "it seems to me" or "it appears to me" or "I feel." When we make objective statements, we say "*any* rational being contemplating this phenomenon would see it this way *if attention were directed to it.*" The conditional part of this statement is important. The statement "climate change is occurring" is probably true, because it is epistemically incoherent for rational creatures to deny its truth *if they direct attention to the relevant facts.*

Notice we don't say that it would be *logically self-contradictory* to deny the reality of climate change. But to deny it requires either directing attention away from relevant facts, or proliferating ad hoc hypotheses to make the facts fit an untrue overall worldview. This is a crucial point, because moral statements, like claims about climate change, are not mere tautologies, and thus can't be absolutely proven with purely logical arguments.

Here are three possible types of definitions of "value":

"Subjective" Definitions of Value

(1) "X has value" just means "I value X."

(2) "X has value" means "Some subject somewhere values X."

"Objective" Definition of Value

(3) "X has value" means "*Any* rational being with the capacity for empathy would value X *if attention were directed* to X." That is, "it would be objectively incorrect for someone who is attentive not to value X" (a claim that could turn out to be either true or false, and would need to be argued for).

There would be no logical self-contradiction in *denying* the truth of the third type of statement, any more than to deny the reality of climate change. We could also consistently grant its truth while maintaining that its truth is irrelevant to any action on our own part. The question is whether an *epistemically coherent overall worldview* requires not only granting the probability of some truths in all three definitions of "value," but also acting in accord with them.

Under the two "subjective" definitions, there are obviously plenty of things that have "value." There is no point in quarrelling with this fact. But the more interesting question is whether there are also things that have "value" in the third type of sense.

As mammals with intact social bonding systems (such as Panksepp's CARE and PANIC systems), our capacity for empathy leads us to intrinsically value the well-being of at least *some* selected others (with the possible exception of complete sociopaths). When we say that a friend's well-being will still have value the day after we die, it can be argued that the friend's well-being has value not only because (1) we value it, and not only because (2) there is *someone* who values it, but also because (3) we believe that, hypothetically, *any* valuing being who intrinsically values even a single other person's well-being (as we all do), and has a coherent worldview, *would* be compelled by the requirements of epistemic coherence to value this person's well-being *to the extent that they were to focus attention on it.*

The argument for the truth of this proposition would hinge on the additional idea that our main criterion for intrinsically valuing someone is the extent to which we have come to focus attention on their valu-*ing* dimension. But the sheer *meaning* of such a position (the definition of "value" being used) would involve not merely the idea that I or someone else value the person in question, but rather the idea that they "actually have" value (in sense (3) above), independently of whether I exist or not. Such a claim is not true by definition, but is a claim that *uses* a certain definition, and like any other claim needs to be justified or refuted.

By speaking in this way, we don't say "this definition is better than that one." Nor do we say "I will use this definition and simply ignore all the others." Any term by any definition can be used in a statement, but the real question then arises, whether that statement is true or not.

All moral terms – "value," "good," "ought," etc. – ultimately need to be defined relative to each other. The reason I begin with "value" is that all the other moral terms would become meaningless in the absence of the concept of a subjective being's valuing something. If there were no valuing creatures, there would be no sense in which this or that ought to happen or ought not to happen, or that something is good or bad. Even in those philosophical viewpoints that hold that "the world" or "nature as a whole" could ground values in the absence of *conscious* beings (e.g., Arthur Schopenhauer 1818, or Hans Jonas 1966, or Holmes Ralston 1986), what is being asserted is still that nature as a whole "values" things – and that it would be incorrect for us to deny that such things "have" value. In short, they assert not only that this subjective entity "values" things in senses (1) and (2), but also that *there is* value in sense (3).

The subjective experience of "valuing" in turn also needs to be understood. When I value something, I desire or attempt to achieve something, to perform some action, or sometimes simply to have a certain conscious state (or the absence of some conscious state) as the outcome. To say that I "value" something is to say that I consider it worthy of some action. I "will" something in Schopenhauer's sense. But we shouldn't oversimplify; we will find as we go that the subjective experience of valuing something is more nebulous than it at first seems, even from the standpoint of how it feels phenomenologically – as Gendlin emphasizes.

Despite these difficulties, subjectively "valuing" something is an organismic process that has to be defined ostensively by experiencing "what it's like," just as with other subjective states like pleasure, pain, hotness, or coldness.

People use the word "value" in a moral context because of its relationship to the subjective experience of valuing; but this fact doesn't at all imply that "X has value" just *means by definition* "I value X." That would beg the question in favor of egoistic hedonism. "I value X" is *one* perfectly legitimate meaning of "X has value," but there are also other meanings. It can also mean "X is worthy of some action" – in our sense (3) above, the sense that it would be *incorrect* in some way not to value X, because not valuing X would render our entire worldview incoherent.

But how can it be "incorrect" not to "*feel* a certain way?" The reason is that, in the third and more interesting sense of "X has value," the attribution of value to something is no longer merely a feeling. It is an acknowledgement of a situation created *by* a combination of feelings – a situation in which *someone* creates what we call "value" by valuing something, such that any *other* valuing subject would feel the value of this valuing creature *if attention were appropriately directed.* I want to argue that it would be wrong or incorrect (because incoherent) for creatures like us *not* to subjectively value certain aspects of this situation in which someone creates value by valuing something. This claim is not and should not be packed into the mere *definition* of "value." If it were, it would beg the question. On the contrary, the claim should be treated as a substantive claim to be argued for, using words that have been defined. The arguments must come after the definition is in place.

Consider a concrete example of "value" in sense (3): When there is an oil spill, we can negatively value the situation because it has destroyed all those valuing creatures (birds, dolphins, etc.) that otherwise would have been in the process of creating value by valuing things. In short, we can adopt the opinion that the creatures themselves have value because they are in the situation in which they create value by valuing. Then derivatively, we can negatively value the situation because it has created suffering for those valuing creatures that still are in the process of valuing.

This approach doesn't lead to the absurd conclusion that there should be a maximum number of valuing creatures on the earth, which would create horrible overpopulation. All the reasons why overpopulation would be undesirable are precisely the countervailing values that oppose the prima facie value of too large a number of valuing creatures.

In all these cases, we can speak in terms of "prima facie" values that can come into conflict with each other, and must be balanced in some way. And of course all values initially have a prima facie status, until we decide what balance between them is best. In some cases, Mother Nature has already made a determination. The well-being of tigers is often in conflict with the well-being of some other animals, and vice versa.

Even if it were not the case that it is better for there to be valuing creatures than for there not to be (that in the valuational realm there should be "something rather than nothing"), the question would still remain whether there are values in our sense (3). After a valuing creature actually does exist, would the attitude be sustainable within a coherent worldview that the values of the valuing creatures have no sense (3) value whatever, even prima facie?

The still more difficult question for our purposes is not whether there actually is such a thing as "value" in any or all of the senses defined here. A harder question is why we should *care* about those values. Why should I value the situation in which some valuing creature *other than myself* creates "value," even in sense (3)? This, again, is a substantive question that cannot and should not be resolved by definition. It is an argument to be made, and the argument should not be a mere tautology.

4.2 Some Crucial Coherence Considerations

As creatures who must act, we can't take the position that nothing matters – that we value nothing. Whatever meaning we then attach to the phrase "X has value," we find ourselves engaged in a process of weighing the coherence of various combinations of criteria for such statements. For example, what criterion would allow the conjunction of "I value the well-being of my children" with "I am justified in beating my children regularly"? To be sure, those two statements can be logically non-contradictory, but to avoid the contradiction requires further assumptions (for example, "Beating children is good for them"). These additional assumptions in turn raise further questions for the coherence of the person's entire worldview.

Similarly, if we think of ourselves as emotivists, we run into coherence challenges for such a belief system. We find ourselves trying to create a coherent view of the realm of valuational statements such as "My friend's well-being has value only in the sense that I value it" (in sense (1) above) versus "My friend's well-being will still have value even if I die tomorrow" (in sense (3) above). We try to forge a coherent moral worldview out of the totality of these coherence challenges.

Especially, we want to have coherent criteria for how to decide *how much* responsibility we have for *whose* well-being. Some such worldviews hold up better under epistemological scrutiny, and we are somewhat justified in thinking that they are more likely to be true than others. But, just as in natural science, there is no such thing as absolute apodictic certainty in this domain.

Someone might object that it is quite logically coherent to say "I value my friend's well-being, but I don't say that my friend's well-being actually '*has*' value beyond my simply valuing it." But once we get into the realm of constructing an overall worldview, the situation is going to become too complicated for such an attitude. It is true that we learn to understand the *process* through which things come to have value by observing and feeling this process in the case of our own person. We know our own being is worth preserving simply because we value things. But if we take this to mean that all values are dependent on our own subjective conferring of value, then a curious version of the "problem of indexicals" overtakes us:

(a) The Moral "Problem of Indexicals." The traditional version of the problem of indexicals challenges us to explain how a physical description of the universe would be any different if "I" were to occupy Bill's consciousness rather than my own. It seems paradoxical that, on the one hand, if everything in nature is physical, then there should be no difference between two universes whose physical description is the same. But on the other hand, it seems intuitively that a universe in which "I" would occupy Bill's consciousness and body would be very different from this one, even if it were physically exactly the same. Bill's body and brain would be exactly the same as they are now, except that "I" would occupy them. I would simply experience the world from Bill's point of view rather than from Ralph's point of view. The comedian Woody Allen tweaks this intuition with his remark "My only regret in life is that I am not someone else."

In the traditional non-ethical version of the problem of indexicals, if "I" were to occupy Bill's brain and body, this would only mean that *an* "I" would experience Bill's subjective consciousness; but the "I" in question would be *Bill's* "I," not the one I have come to know as mine. The Ralph "I" in that alternative universe would still occupy the Ralph body, just as in our current universe. The supposedly imagined case in which my "I" occupies the Bill brain and body *is not possible* without corresponding changes in the physical world, because the brain and body ground the consciousness and personality of each subjective "I." Yet both are instances of an "I."

In the case of subjective *value* experiences, the problem of indexicals manifests itself in this way: If I were to gradually morph into another person, at what point can I honestly say, *from my current perspective*, that I would no longer have value (or be worth preserving) in that slightly morphed condition? By extension, the same question becomes: If my existence is worth preserving simply by virtue of the fact that I am a valuing creature, would my being still be worth preserving by virtue of the fact that I am a valuing creature if I were inhabiting *someone else's* body and consciousness (as Woody Allen supposedly wished he could do)?

In fact, it can be argued that I can't coherently value my own well-being without acknowledging the value of the different person into which I am *always already* in the process of morphing. A person's being consists of a pattern of change rather than a static collection of inert matter (for example see Ellis 1985, 1996, 2005). An organism is a type of self-organizing process that constantly exchanges energy and matter with its environment while maintaining continuity in the pattern (Kauffman 1993). The notion that we could preserve our own identity *without* morphing into a slightly different being is therefore an incoherent idea in its own right. If I were to remain the same from one moment to the next, I would cease being alive. I would literally cease being a self-organizing system. I am always in the process of morphing into a slightly different person, and cannot value only the being of the infinitesimal "current" version of myself. The notion that *this* version of "myself" that exists in this current infinitesimal instant is the only one worth preserving, or capable of grounding value, faces serious coherence problems.

It might seem that an easy reply is that this slightly different person after the morph would have value only in the sense that he or she would value his or her *own* preservation. In other words, it could be maintained that "X has value" just means "I value X," regardless of which "I" is the subject.

But this reply becomes more difficult when we notice an interrelated problem. The notion that only *my own* subjectivity can ground value seems to entail egoistic hedonism. The egocentric stance – the notion that others have value only in the sense that "I value" them – is complicated by the fact that we do by nature attribute *intrinsic* (not merely instrumental) value to certain selected persons *other than* ourselves (for example, our close friends). This is an inevitable result of the nurturance and social bonding systems in the mammalian brain. If Bill's value depended on my act of valuing, then coherence would require me to acknowledge that his interests would have had no value if I had never existed. But by *intrinsically* valuing his interests – not merely *instrumentally* to my own interests – I take the position that their value is not dependent on my existence. This feature of our attribution of value is difficult to coherently square with the belief that "The reason Bill's interests have value is that this other person, the indexical Ralph Ellis, values Bill." We would have to fit that statement with our commitment that "Bill's interests already had value even before Ralph met Bill." Would we posit that Bill's interests had value simply because at some future time Ralph *was going* to meet Bill? If my position were simply that Bill has value only because I value him, then this value would be by virtue of his being instrumental to my own interests. But as a type of creature hardwired to *intrinsically* value at least certain specific others – not merely instrumentally – I can't very coherently take this position.

I don't say that such a position is self-contradictory. But it will become increasingly obvious as we proceed that it faces many coherence challenges. A more coherent viewpoint is simply to say that, by gradually learning to empathize with Bill, I have learned that the things valued by him come to have (prima facie) value *by means of the same process* through which the things valued by Ralph come to have (prima facie) value. It is not merely that "I value" his well-being. Instead, by positing his value as intrinsic, rather than instrumental to my own well-being – by positing his value as not dependent on my existence – I posit that I *or anyone else* who came to empathize with him in the way that I do would also intrinsically value his well-being, unless they were to introduce incoherencies into their overall value system in order to deny his intrinsic prima facie value. If my intrinsically valuing him correlates simply with *paying attention* to his valuing dimension – rather than anything specific to our particular type of relationship (blood relative, adoptive child, comrade in arms, etc.) – then this would imply that anyone else paying *the same* attention would intrinsically (not just instrumentally) value his valuing dimension in the same way.

For this reason, the idea that Bill's being has no value in any other sense than "I value Bill" would be difficult to square coherently with the position that his being has intrinsic value, not merely instrumental value *relative to my own interests.* The more I come to empathize with Bill as I get to know him, the more I understand that his interests have value in the same way that my own changing values continue to have value as I gradually morph from one version of myself to another. The difference between me and my morphed selves is in principle similar to the difference between me and any other valuing person.

The more coherent way to assess this situation is that the value of a subject's well-being doesn't depend on *which* subject the being is. It makes no difference whether the subjectivity that grounds value is the subjectivity indexed to Ralph or to some other subject. Either way, the interests of the subjective creature "have" value in our sense (3) above, at least prima facie. It isn't simply that Bill's values are valuable "to Bill" and Ralph's values are valuable "to Ralph." That view is difficult to reconcile with the problem of indexicals combined with the impossibility of a true egoistic hedonism in which all other values would be contingent on my own interests. When we notice that nothing more than *focusing extensive amounts of attention* on Bill's valuing dimension is what leads us to take the position that his well-being has intrinsic rather than merely instrumental value, we can't coherently avoid positing that any other creature capable of both empathy and overall epistemic coherence *would* value Bill's well-being *if the same attention were directed to it* that we ourselves have directed. This is why people routinely make fun of the incoherence of the anti-Semite who protests, "Some of my best friends are Jews!" Everyone understands the incoherence of thinking that directing increasing amounts of our own attention to someone is what *causes* him to have intrinsic value.

In the analogous case of global warming, we say that it is "objectively true" that global warming is occurring because any rational creature would have to agree that this statement is true *if attention were directed* to the relevant perceptions and measurements. In light of the problem of indexicals and the fact that we don't value Bill's well-being only instrumentally, but rather intrinsically, it is difficult to coherently maintain that Bill's interests would have had no value if we had never met him. To be sure, there are ways to argue against this point, and those will be explored further as we proceed.

(b) Coherence and the Value of Valuing Creatures. The idea that ethical statements can be true or false independently of my own individual subjectivity is a difficult argument to make, and will be built step by step as we proceed. Ultimately, I hope to show still more evidence and argument that we can't coherently maintain that the determinative criterion for how much we value someone's well-being is simply whether they are our blood relative, or in a sexual relationship with us, or members of the same military squad. Those conditions are not always present whenever we come to intrinsically value another person. Instead, we acknowledge the value of valuing creatures to the extent that we *focus our attention* on the valu-*ing* component of their existence. It is true that other factors such as blood relationship or military comradeship do *indirectly* lead us to value a person's well-being because, when we are in those relationships, we do focus more attention on the valuing component of a person's being. But we value it not *by virtue of* the blood relationship, military comradeship, etc., but rather because those relationships tend to *focus our attention on their valuing dimension.*

So there is a coherence problem if we try to think that the person's relationship to us is what *creates* the value that we experience in them or attribute to them. That simply is not what determines whom we intrinsically value. Instead, the consistent pattern is that we intrinsically value others when their relationship to us allows us to *focus attention* on their valuing dimension. We therefore take the well-being of those individuals to *have had* value even before we met them – as phenomenologists like Scheler emphasize. Like the contents of a refrigerator, it was there before we opened the door, and will still be there after we close the door.

As a result, the comparatively more coherent worldviews will attempt to describe not merely what we value, but rather what we take to "*have* value" in sense (3) above – what we posit would still have value even if we were to die tomorrow or even if we had never existed. Hans Jonas (1966/2001) extends this notion by arguing further that life per se creates value (including non-human life), and that it is up to us to consider what is the appropriate balance within our own value system of all these values within the realm of life. Later I will go more precisely into the relationship between these two concepts – our own feelings of value on the one hand, and on the other hand the belief that something "has value" independently of ourselves.

It is important to bear in mind that such arguments shouldn't pretend to "prove" their conclusions foundationally. In my view, the best we can do in moral reasoning is to show that one viewpoint is more likely to be true than competing viewpoints because it meets a number of coherence requirements that increase the likelihood of a viewpoint's being true. We can compare the relative coherence of thinking that people have value by virtue of our specific types of relationships to them versus the view that we take them to have value by virtue of their being valuing creatures. In the latter case, their relationship to us is the *mode through which* we focus increasing amounts of attention on the valuing aspect of their being. It is not self-contradictory to deny this view, but I want to argue that such a denial is considerably less coherent than the alternative.

Ethical viewpoints are never apodictically certain. But they can hope to be more coherent than alternative views. Later I will outline a sense of "coherence" designed to increase the odds of forming overall viewpoints that are closer to the truth. Accordingly, we can expect no more than to show that the view just sketched out is more coherent than competing views such as the purely emotivist perspective. At the same time, it will become obvious that "coherence" requires much more than the mere avoidance of internal logical self-contradiction. Almost any moral viewpoint can be made logically consistent if someone is willing to self-deceptively reinterpret countervailing considerations, as Nazis, racists, and many others have done throughout history.

Emotivism actually would have a good bit in common with the view just described, because it could be constrained by the truth-seeking motive in the sense that, as Salmela (2014) elaborates, we strive for "authenticity" in our emotional lives, to avoid error or self-deception as to the meaning-references of what we actually feel. And authenticity in this sense would also require what we are calling "coherence" of overall worldview. More on Salmela's point later.

For now, the point is that the crucial *criterion* for whom we value intrinsically is the extent to which we focus on *their valuing dimension.* As a result, it is difficult to avoid the implication that Bill has value in sense (3) above – the sense in which any rational creature who intrinsically values even a single other person would also have to recognize Bill's intrinsic value if they were to pay attention to it. Otherwise, we would have to claim that there is some *other* criterion for whom we do or don't intrinsically value.

A related objection against this point might run as follows: Suppose paying attention to someone's valuing dimension were *not* the criterion. Suppose it were to turn out that we consistently intrinsically value others' well-being to the extent that they look pretty, for example. Would this mean that only those who look pretty have intrinsic value?

The crucial problem with this objection becomes apparent when we acknowledge that looking pretty is an instrumental rather than intrinsic value. If I value someone because they look pretty, I value them because their good looks please *me* or because they please *other people* – i.e., the good looks are valued instrumentally, not intrinsically. The intrinsic value in this case is my own pleasure in looking at someone who looks pretty, or the pleasure of other people in looking at them.

If looking pretty were a prerequisite for my intrinsically valuing someone, this would mean only that I am incapable of focusing attention on the valuing dimension of a creature unless it looks pretty. The looking pretty dimension would be only an *indirect* cause of my *intrinsically* valuing the creature – not the aspect that I actually intrinsically value. The creature's valu-*ing* dimension is what actually causes me to *intrinsically* value the creature.

Similarly, if I were to value people only to the extent that they are diligent workers, I would be valuing them instrumentally. And the same if I value them only to the extent that they are good musicians or good athletes. The only plausible candidate for a criterion by which to *intrinsically* value people is the extent to which I focus attention on their valu-*ing* dimension. Someone's valuing dimension is the only valuable dimension of them that would not become an instrumental value in relation to my own subjectivity or the subjectivity of others. It is the only dimension that I posit would have to have been valued by any other human capable of empathy, on pain of incoherence in their overall value system. And again, I am *hardwired* to acknowledge this intrinsic value for at least *some* specific people (for example, a mate, my children, or my military comrades), simply by virtue of focusing so much on their valuing dimension. The question is whether coherence requires acknowledging it for all.

I don't pretend that there are not possible emotivist answers to this argument, and the comparative coherence of those will be considered in due course. But when we examine the criteria for whose valuations we intrinsically (not just instrumentally) value, it becomes increasingly likely that in the final analysis there is no coherent justification for excluding *anyone* from this category. In short, to index the attribution of value either to one particular subject (myself), or to an arbitrarily selected group of specific subjects (for example, my family and tribe), is not the most coherent accounting we can give of the overall interpersonal situation. We can intrinsically value the valuing dimension of an adopted child just as well as we can our genetic offspring, provided that we have come to focus attention on the child's valuing dimension.

(c) Emotivism versus Realism. There is still another important objection to my view as just outlined. Why should it be more coherent to describe ethical "beliefs" as beliefs at all – as if they were the kind of statements that are *capable of being true or false* per se, rather than just *expressions of emotion?* Neither of these two viewpoints is completely epistemologically incoherent. Even a sophisticated *emotivism* could demand that coherence in one's overall value system is preferable to incoherence because of the importance of emotional authenticity and self-honesty.

Mikko Salmela (2014), for one, has done an admirable job of clarifying a notion of emotional authenticity that could be helpful in such an endeavor. Salmela labels as "false emotions" any emotional self-reports that misrepresent what the emotion is "about" in the world. I may say that I am angry at my wife, when actually I suffer from a vague, nagging feeling that I have committed myself to the wrong career decision. Such misperceptions may occur because of self-deceptive processes, or unwarranted popular misconceptions about what feelings are "supposed to" mean, or due to some other mental condition preventing us from focusing effectively on our feelings. Salmela argues that emotions can be "true or false" in the sense that an emotion is (in part) a sensing of a relation to the environment, and may represent or misrepresent such relations.

To his credit, Salmela doesn't say that emotions are always *reactions to* the environment. They could be felt relative to a need to act as well as react. In that respect too, an emotivism that satisfies Salmela's "truthfulness" criterion could be a somewhat coherent position, consistent with what we now know about emotional neuropsychology.

Salmela doesn't take a position with regard to emotivism in ethics, but his analysis shows that there can be both more coherent and less coherent moral viewpoints even within an emotivist framework. I may disapprove of a public figure's sexual misconduct, unless the behavior is done by my own preferred politician; thus my "feelings" would be self-contradictory in a hypocritical sense, hence incoherent. I would be claiming to feel what I don't really feel.

Despite the feasibility of preserving an emotivist framework, I want to argue that it is more coherent on the whole to think of ethical statements as statements capable of being true or false independently of the person who utters the statements. If value statements in our sense (3) can be true, their truth doesn't depend on the feelings of any particular subject. And this view is at least as consistent with Salmela's analysis as is pure emotivism.

The moral realism versus emotivism question hinges on what it means for a reality to be independent from us as opposed to being amenable to our arbitrary whims. The self-organizational approach to consciousness and emotion, especially in its "enactive" articulation, is surprisingly helpful with this question. Suppose it is true that we understand the external world by attempting to *act*, and then observing which actions are *resisted* by the world. In that case, as Newton (1996) points out, the systematic *inability* to do certain actions relative to certain environmental conditions tells us something about the independence of those conditions from our own subjectivity. Similarly, many phenomenologists have noted that statements are believed to have truth value as claims about real facts in the world if we can't alter their truth by means of an act of subjective will (for example, Husserl 1900; Scheler 1954/1970).

In the case of ethical thinking, the question arises: Are there certain combinations of value statements that – much like a rock blocking my path – *resist* my attempts to integrate them within an epistemologically coherent value system? I can't by any act of will induce myself to embrace *knowingly and honestly* an extremely incoherent value system (for example, one that I recognize is logically self-contradictory), any more than I can by fiat will away a stone that gets in the way of my moving foot, or will myself to honestly believe that $2 + 2 = 5$ (though I may pretend to believe so).

Many of these "resistances" in ethical thinking will be explored later in this book, but the following section presents a sketchy preliminary indication of the direction in which our reasoning needs to go on this realism versus emotivism point.

4.4 Summary of the Overall Coherence Argument

A coherent overall moral worldview requires integrating in a consistent way a number of diverse facts relating to creatures that, like us, must act, thus must value, and also have basic CARE and other social-bonding systems in our brains. Integrating all these facts coherently is not as easy as one might at first suspect.

(1) Because of the endogenous and innate nature of certain now widely recognized social emotions, *it is not coherent to claim that I don't intrinsically value anyone else's well-being.* A possible exception is the case of the rare "true sociopath" who literally has no experience of moral guilt. But even if I were such a sociopath, assuming an intact intellect, it would still be *intellectually* evident to me, by examining both neurophysiological evidence and logical reasoning, that if someone *did* have the ability to empathize, there would be certain *criteria* by which empathizing with some while refusing to empathize with others would be inconsistent or incoherent in various situations. Even a high-functioning sociopath could then notice the pattern of these criteria. There is a sense in which relatively high-functioning sociopaths "understand" moral principles, but can't induce themselves to *care* about them. By contrast, a *low*-functioning true sociopath may interpret someone's statement that "X is morally wrong" to mean merely that that person is simply very *displeased* with X. In this respect, *high*-functioning sociopaths often may feel that one "ought" to care about others' well-being, yet because of a peculiar lack of empathic attention they don't emotionally value it enough to affect their own actions.

I freely admit that if humans were completely *incapable* of such emotions as sympathy and compassion, then no such thing as ethical behavior or even understanding would be possible, just as it does no good to preach the well-being of mice to a cat. With no empathic capacity whatever, we would all be extreme sociopaths. But the converse is not true; empathic emotions by themselves can't establish a coherent moral viewpoint.

Correlatively, if it turns out after more empirical research that sociopaths are simply *incapable* of moral behavior, and thus not to be "blamed" for their moral shortcomings, there is no need to worry that this would mean not punishing sociopathic criminals for their crimes. They would still be punished, as a means of deterring not only them, but also other potential offenders.

The first point, then, is that (other than extreme sociopaths) we can't escape intrinsically valuing the well-being of at least *some* others. The problem is how this can extend to a *universalized* compassion, complete with principles of fairness and impartiality.

(2) Careful phenomenology shows that the most important criterion for whether I intrinsically value another's well-being is the extent to which I

focus attention on the *"what it's like"* dimension of the other's consciousness – and especially on their *"what it's like to value something"* dimension. Therefore I know that I *would* intrinsically value (at least prima facie) the well-being of *any* person to the extent that I were to focus attention on their "what it's like to value something" dimension. Here again, the sociopath fails to understand this point only by having severely distorted empathic capacity per se. In a more limited way, of course, we all sometimes deliberately misrepresent the "what it's like" aspect of others' subjectivity when it suits our purposes. The point for now is that when we do value someone else's interests intrinsically is when we do pay attention to their "what it's like to value something" aspect.

(3) From point (2) we can infer the simpler point that, when and if I do intrinsically value a being's well-being, *I do so to the extent that I focus on the fact that the being in question values things.* I focus attention on the valuing dimension of the person's consciousness.

(4) I can construct a coherent moral worldview based on understanding "X has value" (not merely "I value X") relative to the fact that *"X is a valuing being."* The case in which X has value because X is valued *by* a valuing being then becomes the more indirect and derivative case. For example, an ice cream cone can have value if it is valued by a valuing creature, because the well-being of the valuing creature itself has intrinsic value.

By contrast, I *cannot* construct nearly as coherent a value system based on any other understanding of the nature of sense (3) value – for example, (i) "Value = commanded by the gods" or (ii) "Only my own family or tribe members have value." Example (i) is epistemologically unsupportable and can be contradicted with equal credulity; and example (ii) is inconsistent with a coherent working out of the implications of the psychological points preliminarily outlined above. That is, when I ask myself about the criteria that determine the extent of my valuation of others, I can't truthfully say that it comes down to whether they are blood relations; a more coherent criterion is the extent to which I have understood and viewed them as valuing beings. This pattern obtains whether I have *gradually* gotten to know them as friends or relatives, or whether I am suddenly inspired to rush in front of a moving train to save a stranger's life. If this viewing of others as valuing creatures is the crucial criterion for my valuing of them, then to work out a value system that doesn't account for this fact – for example, a system in which their value is dependent on my arbitrary emotional whims – will present formidable coherence problems.

But within the category of value systems that count everyone's interests as having intrinsic prima facie value, there can be an infinite number of possible ways to balance these various intrinsic values. There can be many competing principles with which to describe the best patterning of different people's responsibilities toward each other, the most just distribution systems for the relevant opportunities and resources, etc. In these respects, the details of reaching a final assessment of which value system is more coherent than others, and who should assume which responsibilities, can't be accomplished in the space of one book, and are well beyond the scope of this one. To justify a negative universal (i.e., "no other value system is more coherent than XYZ") requires extensive work that in principle is subject to revision by later analyses, just as science is subject to revision.

Just as in science, it will always be natural at first to resist the needed alterations in our understanding, and the first tendency will usually be to engage in what Thomas Kuhn (1964) calls "normal puzzle solving" – that is, to accommodate discrepant facts within previously established viewpoints. Correlatively, there will also be need for occasional "moral revolutions" analogous to Kuhn's "scientific revolutions." With regard to the details of working out what is the most coherent moral worldview that we can achieve at a given time, continuing work by a community of scholars, thinkers, and social activists will always be required. And hopefully there will be a diversity of voices, so that we can see and point out each other's hidden assumptions.

(5) It follows from the first four points that to say "I have value by virtue of being a valuing creature, whereas someone else does not have value by virtue of being a valuing creature," presents serious coherence challenges. This coherence question requires confronting the fundamental epistemological problem: Is it more appropriate to regard a coherent moral worldview merely as (a) a coherent accounting of one's own "moral emotions" within an authentic, carefully self-reflective version of *emotivism* (epistemologically similar to A. J. Ayer's emotivism, but tempered by Salmela's idea of "emotional truth")? Or on the contrary, is it more coherent to regard "X has value" as (b) a statement that I believe to be true in a non-subjective sense? By non-subjective, I mean that there are reasons to believe that the statement could have been true even if I as an individual had never existed – as in the "sense (3) value" defined earlier.

To choose between these last two options requires pursuing the implications of the points enumerated above. If I define "having value" in terms of the idea that my own being and my own valuations have value, the implication is that my being is worth preserving simply because I value things. I would then say that whatever I value has value (but only prima facie value) *by virtue of the fact that I am a valuing being.* But it implies more basically that I myself have value by virtue of the fact that I am a valuing being, i.e., because I am a valuing being, it matters whether I achieve whatever I value or not, at least prima facie (all else being equal).

For example, the sheer fact that I am a valuing being is enough to justify my struggling for survival. To have value means to be worthy of taking some action toward the value. I don't need to justify my attempts to survive by any other reasons than that. Conversely, if I were not a valuing being, I would simply give up; I would feel that it made no difference whether I survive or not. The question, then, is whether I can coherently say that my being is worth maintaining by virtue of the fact that I am a valuing being, whereas if I were a slightly different person (the person into which I am always in process of morphing) my being would *not* be worth maintaining (at least prima facie) simply by virtue of being a valuing creature. But if this slightly different person must have value by virtue of being a valuing creature, then so would a still more different person, extending ultimately to all valuing creatures.

In terms of indexicals, the question is whether my being has value by virtue of the fact that I value things, or on the contrary whether my being has value by virtue of the fact that I value things combined with the fact that I am *this particular* being indexed to "Ralph Ellis." This question revolves around the problem of indexicals.

Strictly speaking, to resolve the issue of moral coherence requires exploring some specific criteria for the coherence of beliefs in general. But before delving into the psychological and epistemological details involved in coherent versus less coherent belief formation, it might be appropriate at this point to consider still another burning question that must be dealt with if any of these philosophical considerations are even to be relevant (and therefore, ultimately, non-trivial): Even if coherence favors taking a "moral realist" perspective, and even if we are motivated by the independent exploratory emotion system to find the most coherent possible moral belief system, the question remains how we are to be motivated to *act* according to such moral beliefs. Why should I care whether values in sense (3) are promoted – "values" in the sense that, hypothetically, any rational creature would value X if attention were appropriately directed? This hypothetical statement is very abstract in nature, so why should anyone be motivated to act according to such a merely hypothetical statement, even if it is a true one? Are we not back to the "ought implies can" problem after all? How can we be motivated to act according to "values" merely in sense (3)? In short, "Why be moral?"

4.5 Enactive Commitments, Inspiration, and the Motivation to Act

Even if we affirm that valuing creatures have intrinsic value, we still have to confront the "Why be moral?" question. What motivates me to promote the well-being of a valuing creature, even if we assume that I believe it to "have value" in our sense (3)? Of course the specific individuals toward whom I subjectively extend valuing feelings can inspire action, but what is to motivate action toward those whom I merely *abstractly acknowledge* to "have" value in an objective rather than subjective sense – in the merely counterfactual sense that I *would* value them *if* I were to direct more attention to their valuing dimension? It is not altogether obvious how a mere counterfactual conditional statement could motivate action.

As creatures who must act in order to be what we are, it helps to look at our existential constitution to see how we could be motivated to act in certain ways that accord with, rather than conflicting with, the values that we have posited in the most coherent worldview we are able to achieve. One way to think of this self-regulation of our action is to say that the "love" part of our "love of truth" (as Hume called it) motivates us to act in accord with what we have come to believe is likely to be true. If the "love of truth" makes us want to find the truth, it also tends to make us want to "act out" our respect for truth through our actions. But our question now is why this should be the case.

Another perspective on this question is to say that, in order to be conscious of the world in a certain way, we must act in ways that *concretely symbolize* those forms of consciousness. Eugene Gendlin, who worked for many years with Carl Rogers on techniques of emotional communication in psychotherapy, develops this point in *Experiencing and the Creation of Meaning* (1962/1997). Here Gendlin lays out his theoretical orientation in more detail than in the more popular book *Focusing* (1978/1982). In the earlier, more theoretical work, Gendlin shows that the notion that I "love" someone, for example, can become a meaningless abstraction if I never engage in any relevant concrete action to "symbolize" the feeling – to give it an embodied physical expression of some kind, if only through symbolic behavior such as verbal or musical expression. As inherently embodied creatures, we feel things *by means of* engaging in embodied actions, including but not limited to the act of overtly expressing emotions in ways that accord with bodily responses, as in psychotherapy.

As another example, the feelings of sadness that a friend has died can become inert or numb until we sing or play a certain song or say a eulogy, as a physical embodiment through which the feeling can be felt. Sometimes when we speak out loud of a departed person, we are suddenly moved to tears in a way that we were not when thinking the same thoughts silently. Even when we draw or imagine drawing a long division mark and "carrying the remainder," our consciousness of the mathematical operation becomes more vivid (and thus often more accurate).

In sum, such "symbolizing" physical embodiments serve to *intensify* various forms of consciousness. In the present context, Hume's "love of truth" too, which we now explain in terms of the independent SEEKING system and the associated exploratory drive, must be embodied in action. Gendlin's point here is consistent with the enactive nature of consciousness as described by ecological and "enactivist" consciousness theorists such as J. J. Gibson (1986), Ulrich Neisser (1976), and Natika Newton (1996). If consciousness is embodied, then our bodies (including our brains) must act in some way to provide a substrate for the consciousness.

So the compunction to *act* in accord with our moral *beliefs* is felt because we need to physically manifest our love of truth in order to feel it. Feelings are not supernatural things. They need to be embodied through action in order to exist. For similar reasons, when we have acted immorally, our bodies indicate the *failure* to have acted in accord with what we believe to have value (sometimes called "pangs of conscience"). But if we continue failing to act in accord with the values we love (again, because of our love of truth as motivated by the exploratory drive), we eventually *quit* loving them to much of an extent – even if we habitually claim that we do by using stock category words that don't resonate with any bodily feelings. In these cases, the body has achieved a new balance in which what we believe and feel to have value has changed so that the beliefs and feelings are more in accord with the actions or inactions that either provide embodied symbolization for certain values or fail to do so. When we consistently act as if we don't love someone, it eventually becomes a meaningless abstraction to claim that we feel love for them. The situation is the same, I will argue, for the love of truth.

The truth-seeking function and the action-inspiring function of the SEEKING system tend to form a circle with each other. The more we explore valuational issues in a coherent and self-aware way (more difficult than it sounds, as the next chapter will discuss), the more inspired we feel to act according to the resulting values. Conversely, the more we have a feeling of inspiration, the more we want to explore this valuational realm. But the circle can also be a correspondingly *negative* circle. Suppression of the brain's predominantly subcortical inspiration system can also impair the strength with which we are motivated to search for a coherent overall value system, and vice versa.

This *action-initiating* aspect of the SEEKING system is just as important as the exploratory aspect, because we also need a relatively independent motivation to *act* in ways that are consistent with whatever we take to be a moral or valuational "truth." And this will be especially a problem when such actions conflict with other motives such as sheer survival interests, hedonistic gratification, empathic partiality toward an "in-group," or the reinforcements of peer pressure.

In everyday experiential terms, we might not thematize the inner conflict in terms of these psychological constructs. Instead, when we think of moral issues and problems relating to "the meaning of life," we may have a general feeling of being "frozen up," "locked up," or simply "lacking in a feeling of inspiration."

This "locking up" of the energy needed to feel our values as important enough to inspire us is reflected at the conscious level in terms of a basic everyday "existential" conflict. On reflection, we can feel it as a conflict between the need to act in accord with the most coherent overall value system we can achieve, and on the other hand the fear that in the final analysis our entire value system may be "insignificant in the ultimate scheme of things," or perhaps even completely wrong-headed and incoherent. We therefore shy away from venturing out into the potentially inspiring territory, partly because doing so seems hopeless; we fear that it will make no real difference. Sometimes we may also feel threatened by the challenge to our already-existing worldview by asking too many questions about it.

Combined with these existential conflicts are also many neurophysiological issues. The inspiration to act, as we have seen, also requires a physical substratum in terms of neurotransmitter activity and the balance of excitation and inhibition of different areas within the SEEKING system. These physical actions within and on the part of our whole bodies, including our brains, are needed to concretely embody the feelings of inspiration, which again become locked up if there are no strong enough value feelings to fuel them.

The resulting suppression of the desire and/or the ability to focus attention on the concrete experience of intrinsic value through relevant action (by contrast to merely instrumental value) can then create a temptation to narrow our focus toward only the *instrumental* purposes served by our own actions. Lacking an intense feeling of the intrinsic importance of the value of valuing creatures (including ourselves), we seek ways in which we might construe our value (and that of others) as valuable only insofar as it serves some larger purpose, such as earning more money, or solving more difficult math problems, or creating some great technological advance, or building a more successful business. The problem this engenders is that we then tend to treat our own *intrinsic* value (and implicitly the value of others) as if its only "real" value must be instrumental to some extraneous or ulterior value toward which our specific life activities would be merely means. We reduce both ourselves and others to instrumental values – valuable only as means toward some other end. Those merely instrumental values then seem too insignificant to inspire us. "How much difference can one little person's actions make?" we ask ourselves. "How much am I really helping a beggar by giving him two dollars, or five or ten or twenty? And what about all the millions of others?"

Perhaps counterintuitively, this demand for some "ultimate" value toward which we might be instrumental may then make us feel that our own actions are "too small in the ultimate scheme of things," and thus even further to feel a generally depressing *value nihilism.* "Whatever I do in my profession lacks value because many others will do it just as well," we tell ourselves. "If Einstein had not discovered the theory of relativity, someone else inevitably would have discovered it within another five or ten years." Or "Whatever I achieve, it will all be mostly irrelevant by the time I am dead and forgotten."

Notice how such sentiments reduce us to merely instrumental value and ignore our intrinsic value – and by extension, the value of others also comes to seem only instrumental rather than intrinsic. We may then more easily think of someone as valueless if they are "stupid," "incompetent," "wrong-headed," or "worthless." We think of them as lacking in *instrumental* value, and thus fail to focus attention on their *intrinsic* value, which therefore no longer inspires us to action. Ironically, we then feel even more that our actions lack "ultimate" value; serving an ulterior purpose is "not enough," because we are missing the only actual source of intrinsic value – the intensification of our experience of the value of valuing creatures themselves. Now that we have devalued valuing creatures themselves, by focusing only on their instrumental value, we have further exacerbated our movement toward value nihilism. This dynamic can therefore exacerbate depression and the tendency toward nihilism, or alternatively a confabulatory belief system such as authoritarianism, used to mask the underlying nihilism, as Arendt so vividly describes.

4.6 Indexicals and the Egoist–Emotivist Connection

If my having value were by virtue of being indexed to *this particular* being, then I could say that the value of others is only instrumental, and I could be a Hobbesian egoistic hedonist. But with the possible exception of extreme sociopaths, none of us *are* really Hobbesian egoistic hedonists (not even Hobbes himself). As the above observations already suggest, natural emotion systems lead us to value (at least certain specific) others' well-being not merely instrumentally, but rather intrinsically. I can't coherently feel that their value depends on my valuing them. If it did, I would have to regard their value as ceasing on the day I die. There would be no such things as life insurance policies, or even Last Wills and Testaments.

To be sure, A. J. Ayer's emotivism can initially escape this problem because, in his theory, value statements *are neither true nor false* except in the trivial sense that I may misrepresent what I value or don't value. The view could be defended that, if value statements *are merely statements about what I value*, then there actually *is* a coherent explanation for why I might still say that my friend's well-being would have value the day after I die. If the only sense in which I can say something has value is that I, the indexical Ralph Ellis, value it, I could still say that, *at the present time*, I value some *future* state of affairs – for example, the well-being of my surviving friend or of my children and grandchildren. So it is not logically impossible to claim that those future outcomes have value only because I value them now.

But the coherence problems with this view begin to arise when we ask ourselves what we consider to be the *criteria* for whose well-being we do and don't intrinsically value. Both careful phenomenology and mounting neurophysiological evidence show that when we value (at least some) people's well-being intrinsically, this means that we feel that their value is not instrumental to our own well-being. But as we have seen, blood relation or membership in a military squad are not the real criteria for whom we intrinsically value. Paying attention to their valuing dimension is the actual criterion. I must therefore acknowledge that any valuing creature "has value" in the sense that if I were to focus my attention to their valuing dimension, I would intrinsically value their well-being. This counterfactual conditional statement is true irrespective of how I feel now.

If we acknowledge that the term "value" can be used not simply relative to whether I, Ralph Ellis, value something, but rather relative to *whether any given valuing being would value some other valuing being if their attention were appropriately focused*, then there would seem to be a coherence problem with the view that only *my* valuing something, or only *my family and friends* valuing something, can confer value on it.

As already acknowledged, emotivism of the A. J. Ayer kind is not a completely incoherent position, even given the points we have just made. But if I assert a true emotivism, I would still have to admit on phenomenological grounds that "X has value" is not merely and simply equivalent to "I value X." Even in the emotive sense, "I value X" implies "I value X because I can empathize with X's functioning as a valuing being" (for instance, in the case of people who have become my friends, my comrades in arms, my adopted children, or the children of my friends). Yet on the other hand, if I am a true emotivist, I would have to try at the same time to maintain that "X has value" *cannot* be equated to "X is a valuing being" unless I, *this particular* Ralph Ellis person, exist and confer this value upon that valuing being. And I would be trying to maintain this last notion despite the fact that I have already observed in my own phenomenology that I *cannot* coherently say that (at least some) others' value is contingent on my existence, because if their value is intrinsic rather than merely instrumental, then it is not instrumental to my own subjective interests.

So the coherence of emotivism hinges partly on the possibility of a coherent egoism – a system in which the value of others is completely contingent on *my* valuing them, and therefore instrumental to my own emotional concerns rather than an intrinsic value. This leads us to the coherence problems of egoism itself.

4.7 Indexicals and the Coherence Problems of Egoism: Dennett's Alternate Brain in a Vat

To see how incoherent it would be to index all value statements to my own existence, consider as an example Daniel Dennett's amusing thought experiment in his fanciful short story "Where Am I?" (1978). Dennett tries to imagine what would happen if scientists first put Dennett's brain into a vat but left it remotely wired to his body hundreds of miles away. As a backup in case something were to go wrong, the scientists then duplicate Dennett's brain and wire that duplicated brain to a different replicated body (presumably, a hard-wired robotic one). Finally, the original Dennett is provided with a switch so that he can periodically switch places between his two duplicate brains connected to his alternate bodies. When the switch is flipped, brain A is now connected to body B and vice versa. At first, Dennett notices no difference in his sense of self when he flips the switch – only a difference in his perceived physical location and point of view. But the situation becomes more perplexing as the consciousness and personalities of the two Dennetts begin to diverge slightly as a result of their different bodies and different experiences. An even more confusing twist occurs when at a certain point Dennett forgets how many times he has flipped the switch, so now he has no way to ascertain which person is the "real" Dennett and which one is the duplicate.

In the context of ethical thinking, Dennett could have asked: Which one of these people's valuations do I take to be equivalent with a case in which "X has value" – in any of the three senses defined above? When the switch is flipped, and the slightly different Dennett is now the subject of experience, do the original Dennett's valuations now have, say, 90 percent of the value that they had when the original Dennett was the subject, by virtue of the fact that the alternate is 90 percent the same as the original Dennett? Or does the original Dennett now have zero valuational status when the slightly different Dennett's switch is flipped? Or does it have the same value because either subject is equally capable of grounding value?

We might formulate this question with a simple Star Trek transporter thought experiment. The way such transporters are supposed to work is that the pattern of your being is replicated in a different set of atoms in the new location, while the original one disintegrates. But suppose something goes wrong with the transporter so that, even though you are "beamed up" to the ship (which means that the pattern of your being is replicated in a set of atoms on the ship), the original version of you fails to be dismantled in the original location. Now there are two "you"s. The question then would arise as to whether the valuations of the original you would lose their valuational status now that the new you has appeared on the ship. And would the value statements uttered by the original you cease to be expressive of a real state of affairs depending on whether the new you does or doesn't also utter them? Can the you on the ship coherently maintain indifference to the valuations of the you on the ground?

One aspect of Dennett's story illustrates this point especially well – his passing comments about the way the original Dennett feels *toward the alternate Dennett.* Each Dennett comes to have a unique empathic feeling toward the other, stemming from the fact that each one can well imagine what his alternate's "what it's like" quality feels like – including an appreciation of what the alternate's valuations feel like. Consistently with what we observed above, one Dennett's valuation of the other correlates with his ability to empathically understand the other as a valuing being. In an especially heartwarming moment, Dennett reveals that, while occupying the alternate Dennett's body, he feels reluctant to have sex with the alternate Dennett's wife. Luckily, he is able to flip the switch to avoid such a scenario. His valuation of the alternate's values is proportional to the degree in which he empathizes with the "what it's like" quality of the alternate's consciousness.

On the other hand, Dennett also has conflicting feelings, resulting from his own personal discomfort at being trapped in this scenario. But this in no way means that he fails to understand the value of the alternate Dennett. The two Dennetts come to have a "love–hate relationship." Ultimately, we are going to have to consider the effects of these kinds of conflicting emotions.

The more we observe our valuational processes, the more we realize that we value others *to the extent that we understand them as valuing creatures.* When we have finally noticed this pattern, it is difficult to un-notice it.

Many soldiers learn this lesson quite dramatically. They notice that it is impossible to forget that all enemies are valuing creatures after having witnessed the death of some particular enemy. At that point, no arbitrary subjective act can will away our positing that, when others have value (at least prima facie), it is by virtue of the extent to which they are valu-*ing* creatures – at least we can't will away this fact without introducing systematic incoherencies into our worldview. We try to tell ourselves that they "deserved" to die, but our respect for truth reminds us that they were merely following orders originating with their political leaders, just as we were. The "desert" maneuver fails to assuage our regret because it is *logically incoherent.*

Conversely, when we can't *by means of a subjective act of will* change the truth as to which value systems are coherent or incoherent, this is when it becomes appropriate to regard such statements as capable of having truth value independently of our subjectivity. Ultimately, this point becomes the crucial recommendation for regarding ethical statements as having truth value (however fallibly) rather than as merely expressions of emotion. To weave a coherent moral worldview around the idea that all value statements are contingent on my own existence becomes increasingly difficult as the details of such a system play themselves out. These difficulties confront me as independent realities that can't be changed by an act of will, just as a rock's resistance confronts me.

As soon as love of truth becomes a dominant motive, we can't will away our own honestly and carefully considered ethical beliefs by fiat, just as we can't by fiat decide not to value our children's well-being. Nor, as Kant emphasized, can we then will away the resulting "pangs of conscience." What makes things interesting is that our other driving emotions often cause us to *try* to will away our pangs of conscience in order to block out truths, or at least probable truths, because of those conflicting emotions and motivations.

The next chapter will explore more specifically the way this problem affects even honest attempts to evaluate the coherence of competing potential moral worldviews. At this point, it becomes important to consider in more detail the effects of the *inevitably* hermeneutical situation and the problem of the "hermeneutic circle."

PART II

Truth-Seeking and the Hermeneutic Circle

CHAPTER 5

"Attention Must Be Paid!"

Hermeneutics and the Demand for Universalization

Economics is not the "dismal science." Ethics is, for two reasons. First, the ubiquitous chorus of economic suffering becomes dismal only relative to generalized or universalized human values. And secondly, the diversity of ethical opinions itself becomes dismal in its intractability.

Even if everyone accepted some non-controversial value premise such as "pain is bad" (as opposed to simply "*my* pain is bad"), nothing of practical ethical consequence would follow, because the fact remains that other prima facie values can potentially *compete* with the principle that "pain is bad." Granted, for a utilitarian, pain can be justified only to prevent a greater *amount* of pain. But not everyone is a utilitarian. Among ethical theories, as among the general public, there are a thousand radically different versions of the non-utilitarian sentiment "I must do my duty even if the sky falls." If we ask ourselves how to prove logically that one value or the other should take priority, intellectual history shows that we quickly reach a stalemate.

One of the most dismal obstacles to agreement is obviously the tendency for moral feelings to be distorted by selective inattention, which allows self-serving interpretations to tweak moral thinking on an "as-needed" basis. But a still more pernicious problem has to do with *natural altruism itself*; our compassion often remains limited to a circle of friends, family, clan, ethnic or religious comrades, or some other group with whom we naturally empathize. Or the altruism may be withheld from certain groups or individuals, based on some ad hoc justification. Some people, we tell ourselves, don't "*deserve*" our compassion.

Natural emotions such as maternal affection or bonding with a comrade-in-arms can easily establish altruistic sentiments that are *not* universalized. But the search for universalized moral principles is more difficult. And even then there remains the question whether we can be strongly enough driven to *act* on such principles when they contradict non-moral "preferences."

Without universalization of moral concern, the same soldier who risks life and limb for a brother soldier could indifferently commit atrocities toward the enemy, and even enemy civilians. The same corporate executive who is such a devoted grandfather or grandmother could casually make choices that destroy the lives of workers and their families, and then lobby legislators for the right to do so still more ruthlessly – not just out of economic necessity, but also in order to obtain a meager salary bonus or to marginally enhance the appearance of one quarter's spread sheet. The good husband and charitable churchgoer, as history shows, could brutally oppress ethnic minorities, women, or workers in distant dictatorships languishing without labor laws or the right to unionize. As soon as such attitudes are then buttressed by the ad hoc hypothesis that "It's a dog eat dog world," we are already involved in the hermeneutic casuistry of trying to maintain the coherence of an overall value system. We then feel compelled to answer when others point out that, in fact, dogs don't eat other dogs.

When I speak of "generalized or universalized" moral concern, what I mean is this: Suppose a switch on my coffee pot is wired so that it causes a drone to bomb the house of a completely unknown family in an unidentified location on the other side of the world. Would I avoid hitting that switch even if it meant going without coffee? If so, I have at least to some extent "universalized" my moral concern. Whether or not this concern must have an *emotional* component (as I will argue it must), we have already seen that ethical questions have to do with the nature, extent, and patterns in which we generalize our concern beyond specific bonds of affection that are hardwired. But we also need to show how it is *emotionally possible* to prioritize the unknown family in the distant drone's sights – yet without positing fanciful, non-naturalistic mental or metaphysical processes (as Kant thought was necessary, for example). And we must also acknowledge that it is literally impossible to emotionally empathize with billions of suffering people and animals all at the same time.

For all these reasons, many people think that "concern" and "empathy" are unnecessarily emotional words in this context. Many utilitarians (for example, Moore 1900; Singer 1981) as well as non-utilitarians (for example, Pojman and Feiser 2005) may say they would refrain from hitting the coffee switch, not because of any *emotional feeling* such as "empathy" or "concern," but rather out of respect for an abstract moral principle.

Singer, for example, thinks that it is sufficient to take "the element of disinterestedness inherent in the idea of justifying one's conduct to society as a whole, and [extend] this into the principle that to be ethical, a decision must give equal weight to the interests of all affected by it" (1981/2011,

p. 100). Granted, Singer's idea of "justifying one's conduct to society as a whole" is a promising starting point for leading in the direction of *one particular kind* of somewhat generally acceptable *social contract* (among many possible ones – for example, Rawlsian "justice as fairness"). But as Rawls continually emphasizes, a social contract is only a *mutually beneficial cooperative agreement*, not a grounding for a set of moral principles. Rawls is at pains to emphasize this limitation of his argument in the opening sections of *A Theory of Justice* (1971/1999), and he stresses it even more in *Political Liberalism* (1993).

Singer's argument hinges on our desire to "justify . . . conduct to society as a whole" (1981/2011, p. 100). But if that were the whole story, then our motivation to do so would seemingly have to derive from the mutually egoistic motivation to create a social contract – whether a formal political one, or an informal social/cultural one. Singer's lengthy discussions of sociobiology never mention the possibility that Hume's "love of truth" itself might be grounded in an independent natural drive.

Moreover, even if our objective *were* to "justify conduct to society as a whole," it is becoming increasingly obvious that logic alone is inadequate to provide foundationalist proofs to nail down what are the morally best policies, actions, or social systems. Ethics courses that limit themselves to foundationalist logical proofs often tend to send armies of moral skeptics or even moral nihilists out into the world.

Singer remains mostly silent on the question as to what *motivates* me as an individual to care about "justifying . . . conduct to society as a whole," unless this motivation is purely egoistic. Ultimately, in a later commentary on his 1981 argument (see the Afterword to the 2011 revised edition of *The Expanding Circle*), Singer in effect backs away from his earlier position on this point, essentially for the reason just mentioned – while admitting that he has no better argument with which to replace it. Here again, we need to go beyond evolutionary "just so" stories. Tangible brain processes, including but not limited to mere altruistic feelings, are needed to move individuals to care about ethics in a non-trivial sense.

Action, moral or otherwise, has to be motivated. This was why Kant (especially in the *Critique of Practical Reason*) struggled so persistently with his understanding of human psychology when he needed to posit that moral action is *not* necessarily consistent with the motivation to maximize our own hedonistic rewards. Something must motivate us even to care what reason has to say about what we ought to do – *especially* in those cases where we could fairly easily get away with not caring – when we have no need to "justify our actions to society."

Less intelligent animals – wolves, for example – are naturally concerned about the well-being of certain specific individuals in specific situations. The non-trivial ethical questions for humans involve how to generalize moral concern beyond those limited situations. Genocide against enemies seems to have been a commonly "natural" practice during a period between about 20,000 and 3,000 years ago (but probably not so much before that – see Steve Olson, 2002). If we are members of a privileged class, it may be "natural" to feel that educational resources should be predominantly reserved for those who can afford them. But we are still motivated to ask ourselves whether this natural non-universalization is consistent with a coherent value system. For the same reason, universalization in this sense is actually the end goal of Singer's main theoretical work – hence the name *The Expanding Circle* (1981).

If this much is granted, then the addition of generalized or "universalized" ethical concern is essentially what makes economics a "dismal science." Economics is dismal because moral concern extends to all the suffering of the world.

Understanding how universalized ethical concern is possible requires moving beyond evolutionary "just so" stories. It isn't sufficient to posit that Mother Nature hardwired us to behave in certain ways, and that it is somehow to our survival benefit that some of these behaviors sometimes happen to resemble what could properly be called "ethical" in a non-trivial sense. We have already seen that "I ought to be concerned about those whom I naturally want to be concerned about" reduces too easily to a special case of "I ought to do whatever I want."

More is involved here than just universalized "empathy." Edith Stein in her classic phenomenological work *On the Problem of Empathy* (1916/1989) suggests that the way we know what it is like for others to feel pain doesn't require us to *logically deduce* what the other's pain feels like. We know that pain is a datum available to any conscious being because of our own subjective, experiential understanding of what it is to be a conscious being. We understand the concept "pain" as part of what it means to be human even though we can't perceive the other's pain – in the same way that we know that every front side has a back side, even though we can't directly perceive the back side or be sure of its details. We know that the "back side" of a living and conscious creature – the side that we can't see from the outside – includes the capacity for pain.

But for our purposes here, we need more than empathy. We need to understand not only how it is possible to know how another feels, but also how it is possible that someone could be motivated to *expand* and ultimately to universalize the realm of conscious beings toward whom not just empathy, but also emotional *concern* is extended – i.e., beyond what is "hardwired" for all mammals. Knowing how the other feels is one thing; caring about it is another. Moral disagreement, as we have seen, doesn't stem from either a lack of empathy or an ability to care, but rather from the different *patterns* in which the empathy and caring are universalized, and from the ways in which patterns of universalization are rationalized. And these patterns of universalization must somehow be emotionally motivated.

One way to understand the universalization of such emotions is that the terms "human being" and "conscious being" are laden with our understanding of their *"affordances"* in the Gibsonian sense. Just as a nipple for an infant "affords sucking" and an object moving rapidly toward us "affords ducking," so the image of a conscious being (to the extent that it is conscious and we focus attention on it) affords sympathy. We may feel superficially that our concern for the well-being of a friend stems from the fact that he is a friend. But then we notice that the person qua conscious being already "afforded" sympathy before we got to know him and enacted the sympathy. This affordance coheres with our above-mentioned sense (3) of "value":

> (3) "X has value" = "Any rational being with the capacity for empathy combined with a coherent worldview would value X *if attention were directed* to X."

The X in question is another valuing creature. With a nod to Kant, such a principle might be called a "counterfactual imperative."

I suggested in the last chapter that the essential reason *why* we universalize this caring about people's intrinsic value is that we value persons by virtue of the fact that they are valuing creatures. Phenomenologically, the crucial variable affecting how strongly we *attribute* intrinsic value to various creatures is simply the extent to which we can *focus attention* on their valuingness.

But again, it would be absurd to think that we can empathize and have emotional concern for everyone at once. And if not, then how can we "universalize" the concern that we already do feel for a few individuals to whom we do happen to pay attention? This brings us to one of the most important determinants of the capacity for universalized concern: The role of attention is going to be crucial here.

5.1 Universalized Moral Concern and Selective Inattention

How does a notion of "universalized empathy" avoid imposing the intolerable requirement that we constantly empathize with the suffering of all the people on earth, as well as the suffering of animals, at least to the extent that we are aware and pay attention to it? Thinking even for a few moments about the millions of humans in areas of drought and famine and war zones, or even the many species of wild animals losing their habitat, or domesticated animals suffering from inhumane treatment, or the thousands of birds covered with oil from oil spills, can induce a feeling of immense hopelessness, which ironically could *de-motivate* us from acting. Furthermore, if our moral concern is grounded in a sense that the well-being of conscious beings generally matters, then focusing attention on everyday natural suffering and death, even when they are *not* caused by any human wrongdoing, can overwhelm us with virtually intolerable feelings of grief or a depressing nihilistic resignation.

This effect is often thought to be one reason for the high suicide rate among psychotherapists. Their empathy requires vicariously carrying the suffering of many people on their shoulders on a daily basis.

What saves us from complete despair normally hinges on our ability to pay attention to the chorus of human and animal suffering only *selectively.* Most of us, even while trying to do what we can to help, must still for the sake of our own sanity learn how *not* to pay constant attention to the continual mass starvation and preventable disease, the slaughter of senseless constant warfare, and the calloused inhumanity toward people by other humans themselves – although we always retain the ability to remember generally that such things exist.

In psychoanalytic terms, the massive suffering is relegated selectively to our *"preconscious"* rather than to our "unconscious": We retain the ability to be aware of it when called for, even though routinely we keep it mostly out of our consciousness. We can enjoy a beer with friends even while knowing "preconsciously" that children are starving all over the globe. If we were to repress our awareness of the suffering – relegate it permanently to our "unconscious" – we would lose our ability to direct attention to it at will.

Ironically, this need for selective inattention obviously also can lead to a *lack* of concern with morality, or even to incoherent moral reasoning. We often develop *untruthful* ways of inducing inattention to suffering, and as those untruthful methods evolve into entire philosophical worldviews, we find ourselves justifying not only *selective inattention* to suffering – which to a great degree is necessary – but also revised assessments of what we should do about it.

For example, we tell ourselves that the downcast may be miserable because of their own moral shortcomings; they may "deserve" their misfortune, and perhaps ought to be pressured to solve their own problems rather than lapsing into attitudes of dependence. It is for their own good that we refrain from helping such people, and in fact, they may not really be suffering at all, if we tweak the facts in the right way. Life never promised anyone a rose garden, we remind ourselves; people need to learn how to keep a stiff upper lip and work to pull themselves up by their bootstraps, as we might fancy ourselves to have done earlier in life. "If you feed a stray cat," we say, "it becomes dependent and loses the ability to make its own way." *There is* enough truth to such claims to make them into quite effective excuses for selective inattention.

After all, there is truth to the point that people need to take responsibility for themselves, otherwise chaos would result. No matter how just a social contract is, it can't remove all the unfair advantages that people normally and inevitably have over each other. So there is some truth to the position that people need to "play the hand they are dealt." The problem is that either this point or the contrary one can be exaggerated, extended, and twisted to suit our hermeneutical purposes, without our even being aware that we are doing so.

Recently several US legislators justified a vote to cut nutritional assistance for hungry people (50 percent of the recipients consisting of children and 20 percent elderly) by persuading themselves that they were "giving them the gift of work" (for example, see The News Herald, June 29, 2013). In the *general worldview* of these legislators, anyone has the ability to pull themselves up by their bootstraps at any time, and there are ample ad hoc hypotheses to discount any contrary evidence. And if there are exceptional cases not explainable by the ad hoc hypotheses, those are only the exceptions that prove the rule. Such a worldview is buttressed by the perfectly reasonable notion that, if people are not required to accept some responsibility for themselves, chaos would result.

But wait! All of us exhibit this tendency; political liberals as well as conservatives tend to ignore or minimize facts that don't easily fit an overall view of reality. No one is immune to this kind of interpretive or "hermeneutical" process in which we always bring with us to the perception or "reading" of any situation an already existing philosophical worldview in terms of which we interpret the facts. If the facts don't fit the worldview, it may be possible *or even reasonable* to question or re-interpret the facts, or simply direct attention away from them.

Was it *unreasonable* for US senators like Bernie Sanders, Russ Feingold, and Max Cleland to reject the evidence that Iraq had biological and chemical weapons, simply because the Bush administration's claims seemed "suspicious" relative to those senators' *pre-existing overall understanding* of how the world works? Granted, it turned out that they were right to reject the evidence (after two of the three had already been voted out of office); but at the time of their vote, they had no tangible, rational reasons to reject the alleged evidence, except that it didn't fit their already-existing understanding of the world.

Sanders, Feingold, and Cleland were neither irrational nor justified by tangible evidence. They were engaging in the *normal hermeneutical process.* This hermeneutical process is not an aberration, but an inevitable feature of how we form beliefs. We are always "interpreting" reality, not merely perceiving it.

The disturbing aspect of this fact about human belief formation is that none of us is self-transparent enough in our own consciousness to know to what extent we are using the process in a less than honest way, because if indeed we are, we may have developed ways of hiding that very fact from ourselves. The lens through which we hermeneutically interpret reality may also distort our perception of *the lens itself* when we view it in the mirror of self-reflection. If I am distorting reality to fit a narcissistic disturbance, for example, my own narcissistic disturbance may prevent me from perceiving this fact about myself. "I am the most humble person I know," as the saying goes.

The ubiquity of this "hermeneutic circle" provides a daunting reason to fear and shrink away from philosophical self-reflection. We may implicitly sense that, like the family in the film *The Others*, who believed their house was haunted by ghosts, we could be in danger of discovering that the supposed "ghosts" are the *real* inhabitants of the house, while *we ourselves* are actually the ghosts. *We* may be the ones who don't see ourselves as we really are, and therefore are oblivious to our own unwarranted presuppositions.

For example, many anti-Vietnam War activists of the 1960s thought it was irrational for young men to disagree with the war, yet join the military "out of a sense of duty." The volunteers were supporting unnecessary infliction of grotesque harm. At the same time, those who volunteered believed they were responsible for bearing their fair share of a burden that resulted from a democratic process; while they may have disagreed with the specific war, they believed the decision to enter the war had been made through democratic election of representatives, and they believed in the system of representative democracy. They also believed that the burden shouldn't fall disproportionately on the poor and working classes to fight the war, regardless of one's personal beliefs about the war. From the view each of these groups, it was the *others* who appeared to be the metaphorical ghosts in the house. Each side may simply have failed to *pay attention* to the considerations that favored the other side's position.

This brings us back to the way the "hermeneutic circle" itself tends to distort the process of universalizing moral concern. Its first effect is to distort our motivated selective attention/inattention, and thus disrupt the process of "expanded empathy" as an emotional process.

The need to ease our disturbing vicarious experience of others' suffering itself may motivate an elaborate set of ethical distortions designed precisely to avoid paying attention to our own moral uncertainty or the facts that call it into question. We sometimes may also manipulate our own selective attention in order to distort our worldview so as to avoid an intolerable amount of universalized compassion, which can become painful, depressing, and also potentially a threat to a previously developed overall worldview. So a natural next question is to explore the effects of selective attention/inattention on the coherence of the hermeneutical process itself.

5.2 Attention and the Problem of Universalization

To "spread" an altruistic sentiment, experientially speaking, requires minimally that we be able to *direct our attention* to the subjective life of various people – to "what it's like" to inhabit their consciousness. We can then direct attention toward or away from it as needed. We don't direct constant attention to every suffering person or animal, but we know hypothetically that we always *could* do so. The unattended compassion is not "repressed," but is always available in principle to being pulled up into consciousness even though we don't constantly do so. We always "know" that there is world hunger, in the same way that we can mention the dimensions of a room without actually being in the process of performing the measurements.

Although literally feeling compassion for millions of suffering people and animals would be intolerably painful, it is also true that unleashing our feelings of compassion in judicious ways can be as *inspiring* as it is painful. And we have seen that the feeling of inspiration is necessary for an acting being. Action requires valuing, so the feeling of the value of other valuing creatures both determines our actions *and energizes us* by inspiring the actions. Again, this feeling of inspiration – not merely happiness or pleasure – is the opposite of clinical depression.

But how can we *universalize* the feeling of compassion, even in a preconscious sense? We have seen plenty of evidence up to this point that we are hardwired to want to help at least *some* people (if only close friends and family) to the extent that we can identify with what it is like to be in their place. But in what sense does a thoughtful exploratory drive actually enable us to *generalize or universalize* our valuation of other valuing creatures? If such universalization is possible, how does it lead to a more coherent belief system? And if so, to what extent must we extend this "hypothetical" empathy, in what ways, and in what patterns?

Suppose we begin with the concrete experience of valuing *specific* others in our environment (mates, comrades in arms, etc.). As we have seen already, we know in principle that all valuing creatures would have the same ability to pull our empathy *if we were* to focus our attention on them qua valuing creatures. Granted, some people's very obnoxiousness or psychiatric problems get in the way of our being able to empathize fully with them – but this is true only to the extent that we remain unable to focus on their valu-*ing* dimension. As Willie Loman's wife insists in Arthur Miller's *Death of a Salesman*, "Attention must be paid!" – even to the valuing dimension of such an initially unsympathetic figure.

Correlatively, voluntary attention versus selective inattention must be a crucial factor in the ability for moral reasoning, as opposed to mere specific sentiments toward specific individuals. The idea that empathy hypothetically *could* be extended to any given valuing creature is highly suggestive of our earlier "sense (3)" or "non-subjective" definition of value: "X has value" can sometimes mean "Any rational being with the capacity for empathy and a coherent overall worldview would value X *if attention were directed* to X." But how are we to understand the process through which we come to extend even this "hypothetical" empathy to all others in order to affirm that they "have" value in this sense?

For instance, should I merely give two dollars to each individual beggar who pulls my sympathy on the expressway ramp, or should I also send money to an overseas charity? Should I divest from companies that pay bribes to corrupt officials in order to use child labor sweat shops in far away countries? Should I take political action to try to get opportunities and resources to be distributed in a more fair way, or on the contrary should I hold people responsible for adapting to their own circumstances and "playing the hand they're dealt"? Each person whose interests are considered in this "fairness" question is someone to whom I *would* direct empathy if attention *were* directed toward their valuing dimension, just as increasing attention toward a friend's valuing dimension has caused me to have concern for the friend. Even though we can't direct constant attention to all valuing creatures, we still need to be *able in principle* to direct our attention appropriately, if we are to universalize altruism in these ways.

Neurophysiologists have learned a good bit about voluntary attention. Non-human animals control their own attention only in limited and preprogrammed ways, because they don't have the prefrontal hardware needed to finely tune the interaction of various neurotransmitters as they regulate the interactions between the prefrontal cortex (needed to formulate questions, regulate voluntary imagery, and think critically) and the anterior cingulate cortex (used for voluntary direction of attention). By contrast, humans do have the capacity to voluntarily self-regulate these interrelations to a much greater extent, and therefore to control the direction of our own attention much more than other animals (for example, see Damasio 1999; Jeannerod 1997; Ellis 1995, 2005).

The voluntary direction of attention is one of those cases where trying to understand humans by analogy to other animals can lead us astray. Earlier we saw that studies such as the seminal one by Lungu et al. (2007) show how the prefrontal cortex – the main difference between the human and other mammalian brains – can regulate the functioning of the emotional brain (for example, the amygdala), largely by shifting neurotransmitters that control attention, which in turn results in exciting or damping down emotions such as fear or anger. The classic example is when we remind ourselves, when climbing in high places, "Don't look down!" This simple example illustrates, in a limited way, how we can somewhat regulate our fear in this situation by controlling our direction of attention.

In terms of everyday behavior, this means that when someone hates people of a certain nation, race or religion, they must either control the direction of attention in such a way that they don't focus on the "humanity" (or, in our context, the "valuingness") of people in those categories – the subjective quality of what it is like to be those people; or they may have their attention-direction controlled by external forces in these regards.

The converse is not true: Directing attention *away* from the "what it's like" aspect of various people is not *sufficient* to lead to hatred or lack of ethical concern. Ethical concern requires only that an appreciation of the "what it's like" quality of the billions of conscious creatures needs to exist *hypothetically* or *preconsciously*, in the sense that we always know that we *could* execute such thoughts and feelings. This inattention is not the same as hatred or lack of concern. Nonetheless, a *systematic refusal* to pay attention to others' "what it's like" dimension seems to be one crucial aspect of feelings of hatred.

This process of hatred can therefore be illuminating toward an understanding of how our voluntary selective attention and inattention patterns, even without our awareness, can affect our moral valuation or devaluation of others. I use hatred as only one example of an anti-empathic voluntary attention process, but an especially clear one. With that in mind, let's look at this process of hatred in more detail.

5.3 Lessons from Hatred: Motivated Attention as a Factor in Hermeneutical Incoherence

The emotion of hatred dramatically illustrates the interaction between epistemological coherence and the emotionally motivated attention process. As an obvious example, many of us can identify with the emotion of hatred for Nazis. In many cases, the cutting off of empathy required by this hatred is partly caused by the Nazis' own immorality. It is difficult to entertain a moral sentiment toward an immoral consciousness, because to empathize with a state of consciousness requires that in some respects we "resonate" with it in some way. Brain studies show that when we understand another person's action, we execute most of the same brain functions that would be necessary in order to perform that same action ourselves (for example, see Bräten 2007; Jeannerod 1997; Newton 1996); Stamenov and Gallese 2002). So it is more difficult to feel sympathetic toward a consciousness that contains many elements that we can't enact in *our own* consciousness.

This is the focus of the much celebrated studies of the "mirror neuron system" in recent years. Gallese et al. (1996), Jeannerod (1997), Wohlschläger and Bekkering (2002), and other neuropsychologists show that this system allows us to understand others' actions by executing, in our own brains, almost the same neural patterns that we would execute if we were performing those same actions ourselves. This kind of research has led to an increased interest in "enactive" theories of consciousness and cognition: If we understand the world by subliminally imagining how we might act in relation to it, then we must understand *others* by subliminally imagining ourselves performing the actions that they perform (Newton 1996). That we understand others by imagining ourselves performing their *actions* explains why the "mirror neuron system" is essentially coextensive with the *motor system* of the frontal brain areas. We vicariously and imaginatively enact the other's actions.

The act of hating, like any other consciousness, requires using the brain in certain patterns. When we hate, we either refrain from focusing (deliberately or unintentionally) on "what it's like" to inhabit the other's valuational subjectivity – what it is actually like *for them* to experience their needs, wants, aspirations and suffering; or we construct a distorted, highly selective notion of what their consciousness is like. In either case, the effect is to avoid vicariously experiencing the world from that person's point of view.

Gallese et al. (1996) and the various authors in Stamenov and Gallese (2002) discuss the neuropsychology of the process of imagining ourselves in the other's place, especially in relation to the "mirror neuron" system in the frontal brain. The mirror neuron system interrelates with the attentional processes of the anterior cingulate cortex, which in turn interrelate in complex ways with emotional processes of the limbic system and deeper in the subcortex.

Stein Bräten (2007, 2013) provides excellent analyses of the behavioral aspects of this empathy versus non-empathy dichotomy, especially in the context of altruistic versus malevolent behavior (2013). Bräten (2013) describes incredibly heroic and self-sacrificing actions on the part of numerous ordinary people in emergency situations, and equally malevolent actions by similar numbers; the diversity of these actions is coherent with Bräten's earlier (2007) assessment of the effects of the mirror neuron system. In all these examples of both extraordinary empathy and shocking malevolence, voluntary control of attention to what the mirror neuron system can present to us – or lack thereof – is a crucial factor.

So in the case of hatred of Nazis, we often have trouble feeling a "moral" emotion toward someone who is "immoral" (from our own point of view) because it is difficult to get moral and immoral forms of consciousness to "resonate" in order for the empathy between them to occur. In everyday terms, it is difficult to *imagine* ourselves "in the Nazi's place." But if we can understand well enough the ways in which factors such as ignorance or early child abuse, combined with misinformation and intimidation by authorities, could have *caused* the person to adopt the mental dysfunctions of Nazism, then we can learn to set to one side or "bracket" the Nazi's immoral sentiments. In other words, we can learn to voluntarily redirect our attention away from the distasteful aspects in order to focus on the Nazi simply as a valuing creature.

There is no need to empathize with *all* of the Nazi's sentiments in order to empathize with the Nazi's general valuing dimension. No matter how much we may have been harmed by Nazis, if we imagine a Nazi prisoner being deprived of food and tortured, how long would it take us to have some sympathy with his suffering and to empathize with his desire to get fed? At this point, we can learn to set aside or bracket the immorality of the Nazi, and focus more on his capacity for suffering, simply qua valuing creature. Some refer to this as a "forgiveness" process. But to "forgive" in this sense is not to exonerate, or even to forego the person's needed punishment for crimes. It is simply to learn how to direct attention away from the objectionable features of the person and to focus on the valuing function of the person's consciousness.

Suppose, after enabling ourselves to focus on the Nazi's general valuing dimension (which as a matter of fact does exist), while focusing attention away from his immoral aspects, we are then confronted with the image of a Nazi being tortured by someone. Now we can more easily appreciate "what it's like" to inhabit the Nazi's consciousness while being tortured, and therefore we can activate our emotional disapproval of the torture. On the other hand, if we are unable to direct our attention to the Nazi's subjectivity because of the dynamic just described, or for some other reason, then we may remain stuck in the hatred of the Nazi and fail to disapprove of the torture. How we are able to relate to and question our own emotional consciousness, in order to direct the focus of our own attention, therefore plays a huge role in the ability to universalize moral sentiments, and thus the coherence or incoherence of the resulting ethical viewpoint.

I had an older African-American graduate student who had grown up during the worst days of brutality toward Blacks in the United States. He maintained that he was incapable of empathizing with certain people – especially White racists. I asked him to imagine spending five minutes watching the White racist being tortured, then ten minutes, then an hour, then days, weeks, months, years. Finally he granted that, sooner or later, there would come a time when he would begin to empathize with even the White racist's suffering.

At the same time, we have already seen that selective inattention to suffering is often necessary and unavoidable. If it is necessary to suspend empathy in order to kill an enemy in war, we sometimes can suspend it, not because we are incapable of feeling compassion for the humanity of an enemy soldier, but more simply because our own sanity requires that we make a *deliberate choice* to temporarily direct our attention away from the enemy's humanity, in the same way that we "don't look down" when climbing in high places. Underneath the selective inattention, we still know preconsciously that there is a "what it's like" quality there. We may still regret killing a valuing creature *all else being equal*, even though other values may motivate us to direct attention away from the empathy. We still know that it would have been much better if we had not been forced to kill someone who is just as much a valuing creature as ourselves.

Years later at a "re-enactment" ceremony, enemies often literally embrace, because they retain the capacity to focus attention on each other's humanity and valuational consciousness when they decide to do so. In fact, having killed each other may actually *heighten* the appreciation of each other's intrinsic value. Perhaps in some distant future, geopolitical leaders will control their own attention processes well enough to find other means than ordering their citizens to kill each other in the first place.

A friend of mine who served in the thick of combat during the Vietnam War told of a particularly heroic enemy soldier who had been killed; the body was lain out in a very respectful way by the very soldiers who had killed him, with the man's wallet opened out across his chest to display photographs of his wife and children. They then held a somber little memorial service to salute the enemy's valor. The soldiers in that platoon knew how to suspend attention to empathic feelings when necessary, but without completely destroying them, even where the enemy was concerned. Even if the unattended conscious content remains preconscious, it is generally accessible for later attentional focus. Phenomenological methods of the kind that Gendlin has developed (building on the thinking of Merleau-Ponty, Scheler, and others) can be crucially helpful in transforming our ability to clarify the meaning of these emotional processes, and thus to enable more coherent moral perspectives.

As creatures with elaborate and sophisticated prefrontal-anterior-cingulate interactions, and therefore with the ability to reflect phenomenologically in subtle ways, we have the capacity, after going through certain thought processes, to direct our attention toward the humanity of even the enemy and the Nazi, and to empathize with that person's valuing dimension and vulnerability to suffering, simply because there is a "what it's like" quality to the valuing aspect of the person's being. Imagine, for example, a psychotherapist charged with the task of treating the psychiatric problems of a former Nazi. Can the therapist engage in the needed empathy with the Nazi, even as the Nazi continues uttering immoral statements with which the therapist can't empathize?

It will become increasingly obvious as we proceed that *the crucial* question for coherence in our overall moral viewpoint is whether we choose or are able to direct our attention toward the subjectivity or "what it's like" quality, the "valuing" nature of any given person, even if we hold this vicarious appreciation in a preconscious status most of the time. Our directing attention away from something doesn't stop it from existing, but it may sometimes lead us to *feel* and *act* as if it doesn't exist.

These attentional processes are relevant to epistemological coherence. To direct attention away from the enemy's humanity is not necessarily incoherent. Nor is it incoherent to refrain from constantly thinking about all the hungry children around the world. In the legal arena, it is not necessarily incoherent to direct attention away from a criminal's suffering as we sentence him to years of prison. But if we purposely and systematically direct our attention away from something that exists, and then adopt a moral viewpoint that operates as if it doesn't exist merely because we are *not motivated to attend to it* – for example, the unnecessary suffering of the convict, or the "collateral damage" of combat, or preventable injustice regarding the poor, or the plight of war refugees – this is not merely an "emotional preference" on our part. It is a source of incoherence in our moral thinking. As such, it subtracts from the overall coherence of our moral standpoint, and counts against the epistemological viability of that standpoint.

5.4 Poor Phenomenology as It Affects the Coherence of Moral Standpoints

We can see that one of the determinants of how much we generalize our moral sentiments is the patterning of both voluntary and involuntary attention. But then we can also pose the question: Do certain incoherencies in the construction of a hermeneutical worldview facilitate the *selective blocking* of the ability to generalize moral sentiments? Does the inability to direct attention function like an ink-block printing system, selectively *blocking* just enough ink from certain areas to result in the precise pattern of the finished image?

If so, then we can raise objections against the resulting distortions of generalized moral sentiments, not because the person's moral sentiments happen to be "different" from our own, but rather because the moral sentiments in question require for their support an incoherent set of hermeneutical thought processes designed to enforce not merely selective inattention, but chronic selective *refusal to engage* in certain attentional processes. This refusal is not a temporary blocking of attention for an immediate practical purpose, such as when a surgeon makes an incision, but rather a permanent inability to *acknowledge the existence* of something that in fact exists. The resulting overall viewpoints are correspondingly lacking in epistemological coherence.

We can't make an absolutely certain judgment as to the degree of coherence or incoherence of a belief system. And even if we could make such judgments with certainty, the coherence or incoherence of the belief system itself would remain a matter of degree. No finite human has an absolutely coherent or incoherent belief system. But the ultimate goal of the truth-seeking motive in ethics, as in most truth-seeking functions, is to be able to say with some confidence that a more coherent system is epistemologically preferable to a less coherent one. So far, we have only roughly defined what is meant by such "coherence," so we will need to further clarify it in the following chapters.

Someone may worry that the most coherent viewpoints might tend to be the *simplest* ones. The problem here would be that simplicity may not be a virtue of moral viewpoints, since moral issues are complex. Coherence theory, superficially, could appear favorable to the "moral clarity" of those who brush aside ethical nuance and the real complexity of the world with simplistic formulas such as "Terrorists are enemies of freedom and thus have no rights," or "The only thing that really matters is to aggregately maximize human happiness."

But a simpler moral viewpoint generally is *not* more coherent than a more complex one, even granting the role of "parsimony of first principles" in the criteria for coherence. Propping up an overly simple belief system often requires a very complicated system of hermeneutical maneuvers – including illogical inferences, denials of reasonable fact, and intentionally selective out-of-touchness with some of one's own feelings and thoughts. Those who hate terrorists so much that they enjoy seeing them tortured (never mind merely *suspected* terrorists) may seem to have a streamlined, elegantly parsimonious worldview. But when we dig into the system of convolutions needed to maintain such a simple viewpoint, we find thought processes that are by no means simple, elegant, or coherent.

5.5 Coherence and the Role of Emotion

I hope it will become obvious that, in real life, people actually use (or at least try to use) a coherence epistemology in forming their ethical beliefs, at least insofar as they are not overwhelmed by the kinds of confabulatory tendencies we discussed earlier. Coherence epistemology means that we would want to select epistemological criteria that have the best chance of leading as far as possible away from false and self-deceptive viewpoints, and in the direction of more veridical ones, but without expecting absolute truth or apodictic certainty. We would need to establish criteria such as logical consistency, a good fit to our honest best current guess as to empirical facts, and minimizing of ungrounded assumptions and ad hoc hypotheses. We would also need to eschew claims based on mere "moral intuition"; everyone knows that claims based only on intuition can be contradicted with equal epistemological coherence, and this fact implies in principle that intuition by itself has no justificatory authority. Everyone knows this, even if sometimes our hermeneutical gymnastics prompt us to pretend otherwise. Everyone knows that, if different people's logically-consistent intuitions contradict *each other*, neither intuition by itself can determine the truth as to which viewpoint is to be preferred.

Perhaps most important, a coherent viewpoint would have to fit with the most careful and critical introspection we can devise of the interplay of different emotional processes that can affect selective attention, and consequently the hermeneutical development of moral worldviews. It isn't sufficient to begin with a "moral intuition," and then rid it of self-contradictory implications; that would be a recipe for casuistry in defense of the prejudices of our childhood. Such a strategy itself should be taken as evidence *against* the coherence of a viewpoint. There is no escaping the need to examine the *emotional* processes that determine whether such prejudices dominate the hermeneutics of the resulting moral worldviews.

The conflicts between the "love of truth" and concern with "us and our own" (including, as Becker suggested, the existential need to feel in control of our destiny) can become so threatening to a person's hermeneutical worldview that it may now be necessary to deconstruct entire emotionally driven ways of thought that have evolved intergenerationally ever since the beginnings of written history and the accumulation of private property and social dominance systems – as Rousseau (1762) and many others have emphasized. Authoritarian parents, for example, tend to inculcate authoritarianism in their progeny (Duriez and Soenens 2009). The Duriez and Soenens data suggest that intergenerational transmission of authoritarianism determines to what extent even *racism* is intergenerationally transmitted.

If the hermeneutical "interpretive reality" process really does play a strong role in the forming of moral worldviews, then focusing on the conflicts and interrelations of these complex emotional processes becomes a crucial element in a coherence approach to ethical thinking and ethical theorizing. Logic is one tool for this process, but logic can also be used in the interest of hermeneutical casuistry.

Interpreting the *meaning* of our own morally relevant sentiments is much more problematic than many people assume. Is the "intuition" that supposedly justifies torturing suspected terrorists really only a desire for vengeance? Is the decision to use aggressive "mob control" against a crowd of peaceful demonstrators who break a curfew really an expression of frustration with one's own rebellious teenagers? Or is it perhaps a result of unresolved feelings about having been "bullied" as a child? "*This* time," we may be telling ourselves, "I'm not giving an inch!" Eugene Gendlin has developed a method of trying to clarify the meaning of these kinds of opaque (because conflict-ridden) affective processes. Gendlin's (1978/1982) "focusing" method originally evolved from experiments on how to improve how well subjects were able to focus on the meaning of their feelings during therapy sessions. It is by no means easy to make this kind of assessment in an accurate way, and carelessness about it is one of the main causes of ethical incoherence.

Did Florida state legislators *really* feel that it was more fair to allow an armed man who "feels threatened" to shoot people by preference over retreating (the now famous "Stand Your Ground" law)? Or did they only feel pressured to bow to elements of an economically and culturally frustrated constituency for whom wielding guns is a symbol of reasserted machismo? Carefully worked out phenomenological methods like Gendlin's are attempts to get us past superficial readings of the affective processes that affect our own hermeneutical perspective. Such phenomenological methods therefore are especially crucial in the moral realm because of the temptation to distort the hermeneutical filtering of our perception of reality in the ways we were discussing above – for example, perceptions consistent with the idea that, in a universe governed by a good (Christian) God, only a Jewish conspiracy could have caused the Bubonic Plague.

Using methods of getting more clear about the meaning of our own feelings and motivations could be combined with the demand for consistency with logic and with empirical reality and minimizing of ungrounded assumptions. We might then be able to develop criteria for coherence even in an admittedly hermeneutical worldview – hermeneutical in the sense that we always interpret new information in the attempt to make it fit an already existing belief system. Such coherence criteria can't pretend to certainty, but there might be certain sets of epistemological criteria that are likely to yield better ways of moral thinking than others.

The overall set of such criteria can be used to evaluate the relative coherence of different value systems and ethical viewpoints from a *truth-functional* point of view just as well as it can from an *emotivist* one. The fact that ethical statements can't be strictly proven beyond doubt doesn't mean that they can't have truth value or that they are only expressions of emotion. The grounding of some ethical viewpoints may still be more coherent than others, similarly to the way some scientific theories are preferable to others.

None of these coherence criteria would need to claim absolute definitiveness. Quantum theory is full of "anomalies" and logical "paradoxes," but hasn't yet been abandoned, because physicists haven't yet found a more coherent theory, even granted the logical contradictions in current theoretical work. No one suggests that quantum theory, despite its incomplete confirmations and its anomalies, is "only an expression of emotion." All branches of science, as Kuhn reminds us, are faced with more or less occasional anomalies, and are perplexed as to how to make some of the observed facts fit the going theory – yet the theory is not automatically rejected because, for the time being, it seems more likely that someone will discover a way to explain *how the anomalous facts fit* than that someone will discover a better theory.

To be sure, there would be a crucial difference between the epistemology of science and that of ethics if it turns out that ethical viewpoints are not even *probabilistically* "falsifiable" in the way that scientific theories presumably are, at least to some extent. But even in ethics there are facts about the need to act, which entails value, and combinations of these facts can make some systems of value less coherent than others. The attempt to construct a coherent valuational worldview "resists us" in a way similar to the way objective physical realities resist our attempts to manipulate or move through them – which is how we know they are real independently of us. Ethical coherence resists us in the same way as the attempt to reconcile a scientific theory with an anomalous finding.

I don't mean with this preliminary definition of coherence epistemology to justify such an approach, but only to point out that it seems to be an additional alternative alongside emotivism, if we give up the strategy of absolutely proving the truth of importantly controversial ethical principles on the basis of a priori logic and/or empirical methods.

In effect, there are essentially two main reasons why *emotions* become especially important in the context of coherence epistemology:

(i) ***Emotion and Hermeneutics***. We have already begun to see that one of the most crucial causes of incoherence in the hermeneutical process of moral thinking is that we misread the phenomenology of our own emotional processes by introspecting too carelessly. We misinterpret what our own emotions are really about. For this reason we are unconsciously motivated to introduce distorting elements into our interpretations of reality, including the reality of moral situations, and in fact *especially* in those situations. Legislators who cut aid to hungry children may remain unaware of the role played in their own psychology by existential issues such as a need to feel that everyone is in control of their own destiny – quite possibly because each of us needs to feel that we are more in control of *our own* overall existential situation than we actually are.

As a simple example of the way emotions unconsciously distort moral thinking, suppose I am somewhat narcissistically disturbed. By definition, this would mean that I secretly wish to measure up to an impossibly grandiose self-image, and sooner or later will come to resent not having achieved it. At this point, professional demagogues on talk radio or cable news present me with ubiquitous complaints about the supposedly perverse social and political system that has thwarted the possibility of living up to my impossible ego ideal. For instance, I am told that immigrants get special consideration in college admissions, and this explains why I was not accepted into Harvard. Now I am tempted not only to adopt a concept of "fairness" that is almost synonymous with what is good for myself, and to direct attention away from what is good for the immigrants involved. I am also tempted to selectively direct attention away from certain empirical facts – for example, that I myself enjoyed the "unfair" advantage of attending a prestigious prep school, and will have family and school-friend connections to help me get good jobs when I graduate. Instead of considering the implications of all these facts, I choose to view the immigrants simply as "unfairly" taking my slot, and eventually to see the entire society as a zero-sum game. At this point, I had better "look out for number one, because no one else is going to." My overall worldview becomes increasingly distorted to fit the notion of "fairness" that has been formed with certain emotional priorities and needs as its background.

Again, this hermeneutic process would not be a crucial problem *if it were possible to construct foundationalist proofs* in ethics. But such absolute proofs are not possible, as the next chapter will explore in a little more detail. Correlatively, we can't avoid the need to deconstruct hermeneutical worldviews – including and especially *our own* – in order to assess their overall coherence.

(ii) *Emotion and Moral Epistemology*. The second reason for the importance of emotion for coherence thinking is more abstract, but equally crucial: The coherence of moral outlooks hinges in large part on how they handle the dangers of *logical* incoherence that result from paradoxes around the need for moral behavior to be psychologically motivated. Serious ethical thought is squeezed between two contrary constraints: The "ought implies can" principle requires that morality must be motivated; but at the same time, the naturalistic fallacy problem highlights the absurdity of the idea that people should do the right thing only because they naturally feel like doing so.

Even if we can *describe* certain behaviors as "good" or "bad" in the sense that they are helpful or harmful to other people, we still haven't made a non-vacuously *prescriptive* statement if someone can't be motivated to act accordingly. The statement would have the same meaning as when we say that a tornado or earthquake is "bad." No such statement can have any effect on the behavior of the tornado or earthquake, and thus is prescriptively trivial. Even if the tornado had consciousness and could understand our description of it as "bad," it would still be incapable of acting differently. And if it were capable of the hermeneutical distortions of authoritarianism, totalitarian conspiracy theories, and other such confabulations, it might even believe that our designation of "bad" is untrue.

To render any moral viewpoint even minimally coherent requires dealing with this "ought implies can" problem. And this in turn requires a careful analysis of the relationship between ethical theory and the human emotions hopefully revealing that there are natural emotions that can make ethics possible, while at the same time not rendering it trivial by reducing ethical statements to statements about whatever enlightened-self-interest cooperative incentives or even altruistic or compassionate feelings we *already happen to possess.* This is why we have focused so extensively on Hume's "love of truth," whose motivational basis can be found in the innate exploratory drive, as articulated in Panksepp's SEEKING system.

The coherence approach as I have characterized it has a good bit in common with emotivism, because so much hinges on the ability to generalize or universalize altruistic *feelings* that must be capable of occurring naturally. The difference is that, in the coherence approach, our ethical principles don't depend on the corresponding altruistic feelings, but rather on an overall assessment of our worldview as guided by the love of truth. Without the ability to empathize and feel moral concern, humans would be as incapable of morality as snakes. But without a love of truth, we would believe whatever makes us feel good.

For that reason, emotivism per se would threaten to neuter ethics by rendering any moral (or immoral) sentiment just as "valid" as any other, since in that case "ethical" comportment would come into play only when and if I happen to feel altruistic. I am arguing that we can do better than a mere emotivism in ethical theory, if we are willing to adopt a coherence rather than a foundationalist epistemology. Since coherence epistemology seldom leaves room for apodictic certainty, we must allow that emotivism is *one possible* ethical outlook, to stand alongside other viewpoints in an overall coherence comparison. We have already begun to see that emotivism itself faces fairly serious difficulties with regard to the epistemological coherence of the theory.

As soon as the exploratory drive comes into play, we are then motivated by more than just altruistic sentiments toward accidentally selected people. We then have a need to ask ourselves about the criteria for prioritizing various people's interests in various ways, and whether some criteria are more coherent than others. Most important, we want to know the truth as to which actions really are best, regardless of whether our *other* emotions (other than the love of truth) would have led us to prefer this or that belief system. This is a claim that can't be made for any other natural emotion or motivational system than the exploratory drive itself. The exploratory drive is the only one that leads us to seek the truth independently of what we would like the truth to be.

I suggested earlier that the value we attach to others correlates with the extent to which we focus on their *valuing* dimension. If this is true, then the more coherent ethical viewpoints tend to be the ones in which what is good for people, or for valuing beings in general, is taken to have value, at least prima facie or "all else being equal." Once this connection between value and valuing is granted, then we can compare the relative degree of coherence of the various moral viewpoints that make room for it.

But the comparison of viewpoints will hinge crucially on their coherence with facts about human emotions. This would include the motivation to act morally even when doing so is genuinely irrelevant to or even against one's own happiness maximization, as we saw in the previous chapter. And it would also include the emotional issues around what is good or bad for humans, as well as the ways that the problem of psychological determinism affects the coherence of different approaches to ethics. For example, distributive theories of justice are more compatible with determinism than are retributive ones – "retributive" in the sense that some people just "deserve" certain rewards or punishments irrespective of any good outcomes produced by the rewards and punishments. The more determinism there is in human development, the less coherent such a concept of "retributive justice" becomes.[1]

The natural exploratory tendency in humans is the main motivation for questioning our own hermeneutical worldview. Lower mammals are also endowed with a strong exploratory instinct, but they lack the mental apparatus to apply this instinct to ethical questions in the way that humans can.

The crucial ethical questions have to do with the extent to which altruistic sentiments capable of motivating action should also be generalized or universalized. Can we be motivated to distribute altruistic concern in the best possible way, and if so, what is the best possible way?

If motivated to do so, I have the ability to distribute my naturally action-motivating altruistic concern more broadly than I now do. For example, I can direct my attention to literary works and artistic works that pull my universalized compassion. The protagonist with whom I empathize is not even a specific real-life person, but even such a *hypothetical* example of a valuing creature can pull my empathy. I can reflect on images of suffering civilians in a war zone or hungry masses in refugee camps, to remind me from time to time of the "what it's like" quality of these people's experience, which can also inspire me to action. We know how to train ourselves to universalize compassion in the same way that we know how to train for a marathon. To a great extent, this is a matter of understanding some of the principles that govern one's own psychological conditioning. We all have this capacity (with the possible exception of extreme "alexithymics," who paradoxically are unable to feel their emotions as emotions, or to recognize their meaning). And we also have the capacity to hold this now-universalized moral concern in a pre-conscious status, rather than constantly suffering vicariously with all the valuing creatures on the planet.

The larger point is that the conflicts between the exploratory motivation and other motives, as we form our necessarily hermeneutical belief systems, must be a central focus for any coherent approach to ethical epistemology. But it is also necessary that any given person feel "inspired" to put forth this effort with regard to his or her life and social situation; the role of this feeling of inspiration, as we have seen, also involves Panksepp's SEEKING system.

I have sketched up to this point an approach that makes moral theory depend in crucial ways on moral psychology. Again, it would be nice if ethical theories could be foundationally proved without thereby becoming empty tautologies, but as will become increasingly clear in our next chapter, they can't.

Theories and viewpoints can, however, be evaluated from a truth-functional perspective, *provided that* (and this is a huge "provided that") we can deal with a crucial dilemma that we haven't yet confronted: How is it that the independence of the exploratory drive allows us to develop a coherent moral epistemology? Can such a coherence strategy actually serve as a serious alternative to either foundational logical proofs or mere emotivism, even though we have granted that the study of emotion itself plays such a pivotal role in determining the epistemological coherence of hermeneutical viewpoints? And can the idea of a coherence epistemology really be fleshed out in a credible way without already presupposing foundationalist epistemological principles? For similar reasons, the idea that we could ever get past the hermeneutic circle and identify the ways our own emotions and preconceptions distort our overall worldview is a hurdle that is difficult to imagine getting over. We must now consider those difficulties in more depth.

NOTE

1 Systematic analysis of the logical contradictions between determinism and retributive justice concepts would take us too far beyond our scope here, but I discussed the issue more extensively in Ellis (1992).

CHAPTER 6

The Coherence of Moral Worldviews
Beyond the Privileging of Nihilism

In the early twenty-first century, the epistemological grounding of ethical beliefs is as precarious as it has ever been. In academic philosophy, this is simply a historical fact, if by epistemology we mean establishing methods for demonstrating the truth of statements. Philosophers now more than ever question whether any rational method, even in principle, could possibly show that one basic moral belief system is the correct one, especially when it comes to the most important questions – questions involving the balance of competing prima facie values. Should maximizing human happiness take priority over individual rights? Should freedom supersede governmental protection from harm, or vice versa? Can a supposed deterrent effect of the death penalty justify the execution of innocent men who consequently are somewhat routinely executed?

Granted, Alisdair MacIntyre's once-celebrated demonstration of the impossibility of logical proofs of ethical claims was fatally flawed, because he merely cherry-picked his favorite invalid arguments to illustrate the impossibility of proving moral conclusions (MacIntyre 1984, ch. 1). But the fact remains that for thousands of years philosophers have failed to reach consensus on basic first principles of moral reasoning. Mill refutes Kant's arguments, and in turn is refuted by Rawls, who only succeeds in stirring up a new hornet's nest of irresolvable controversies. College ethics courses present students with a smorgasbord of mutually contradictory ethical theories, with little in the way of criteria for choosing among them. It is seeming more and more clear, to both philosophers and everyday people, that ethical controversies can't be settled by logic alone.

But the insufficient foundationalist attempts to "prove" ethical principles can point fruitfully toward a more hopeful approach – toward *coherence criteria* for steering closer toward the truth. In fact, such coherence criteria are what we typically use in everyday moral thinking, to the extent that we avoid the confabulatory tendencies discussed in the previous chapters.

Attempts to logically demonstrate the truth of ethical propositions always seem on closer inspection simply to help themselves to controversial and apparently unprovable major premises. It is now a commonplace among philosophers that the kinds of rational decision methods offered by Kant, Mill, or Gewirth depend on controversial presuppositions. When considered as *foundationalist proofs*, such rational methods thus seem inadequate.

The uncomfortable current state of moral epistemology can be understood from both philosophical and social-historical perspectives. Until the nineteenth century, ethical theory attempted to ground itself in traditional foundationalist epistemology – "foundationalist" in the sense that ethical judgments should be grounded in statements whose truth can be clearly demonstrated in some way. Epistemology in those days was mostly defined as the attempt to understand the methods by which the human mind can claim to know that various types of truth claims are justified (for example, through deductive logic or empirical methods), or whether there are domains for which (as Kant argued in the *Critique of Pure Reason*, 1781) the mind is inherently ill-equipped to know what is true (for example, metaphysics).

It was also traditionally acknowledged that there may be some domains in which there is *no truth or falsity at all*, but perhaps only habitual prejudices, including socially or historically relative fluctuations. Throughout the twentieth century especially, it was increasingly feared that ethical judgments *per se* might fall within this category. This was the central question of McIntyre's *After Virtue* (1984), as well as for the postmodernists and cultural relativists, and also many logical empiricists such as A. J. Ayer.

Increasingly throughout the nineteenth and twentieth centuries, philosophers gravitated toward the question: Is there *any* epistemological grounding for moral beliefs at all, or do we merely succumb to attitudes congruent with cultural tradition, peer pressure, or our own psychological idiosyncrasies? After all, even as recently as mid-twentieth century, people variously have held extremely controversial beliefs about homosexuality, freedom of the press, racial inequality, and the rights of workers to form labor unions, which greatly affect whether the balance of power between workers and management is "fair" or "unfair."

People often are not even clear as to whether their "beliefs" about such things actually *are* beliefs, as opposed to emotional dispositions – or even whether the attitudes in question *have anything to do* with morality or not. Introspectively, it is difficult to distinguish between an "intuition" that "homosexuality is morally wrong" and merely an emotional feeling of disgust that may result when heterosexuals imagine homosexual behavior – just as someone who dislikes chicken liver might be disgusted by the image of someone eating it. Thus "common moral sentiments" can be crucially misleading in ethical epistemology – as they often have been used *even in academic philosophy.* In an era of rapid change, it becomes more obvious than ever that these "intuitions" may often or even always represent merely, as Einstein put it, "the prejudices of our childhood."

If we take Ayer's reasoning in *Language, Truth and Logic* (1946) seriously (see also Ayer 1965), we intuitively "believe" that certain things are right or wrong only because we have emotional proclivities toward certain kinds of behavior, as a result of cultural tradition or our own emotional psychology. We then can look only to sociologists to tell us why we have the particular beliefs that we have. On this view, philosophy, as A. C. Danto (1968) put it, has "worked its way out of a job."

Notice, however, a crucial contrary point: *Ayer is assuming a foundationalist epistemology.* His argument is that ethical statements can't be proven by means of logical and/or empirical arguments – and therefore there can be no rational reasons to accept them. He doesn't consider that beliefs may be supported in ways that are not foundationalist – for example, by using coherence methods.

Ayer's resulting relativistic way of thinking about ethics has become increasingly prevalent in philosophy as well as everyday life. In effect, we all bought into Ayer's logical-empiricist arguments – even those of us who literally had never heard of them. Philosophical belief systems tend to filter down from the students of earlier philosophers, to *their* students in various disciplines, including literature, film, and the social sciences. Eventually, the take-home messages spread to everyone, even if in oversimplified form. So establishing what the truth is on ethical questions, or even whether a "truth" exists, is more of a challenge than ever before. Like lingering ghosts, foundationalist assumptions such as Ayer's logical empiricism languish in limbo, refusing to accept that they already have been dead for a century or more.

6.1 What "Intuition" Is Not

The precursors of an alternative approach – a coherence approach – began to develop toward the turn of the twentieth century (Moore 1900; Ross 1930). Some philosophers argued that we can begin with "common moral intuitions" and then gradually "refine" those to rid them of logical self-contradictions. G. E. Moore's *Ethics* (1900) and W. D. Ross's *The Right and the Good* (1930) are classic examples, and more recently Louis Pojman and James Feiser have used this approach to a great extent (for example, in *Ethics: Discovering Right and Wrong*, 2005). We discussed in Chapter 1 the way Norman Daniels (1979) and Michael DePaul (1986) further developed this method. In the philosophy journals, discussions using the method of "refining intuitions to make them logically consistent" now appear routinely.

But as we have already seen, to begin with "intuitions" can be a recipe for casuistry. Even a *non–self-contradictory* "intuition" still clashes with the differing intuitions of other people, especially given a history of different life experiences. If intuitions were an indication of truth, then it should be logically impossible for different people's intuitions to contradict *each other.* True statements can't logically contradict each other. So the real grounds for deciding who is right must be something *other than* intuition. The intuition alone has no justificatory force.[1]

What, then, of the obvious fact that in real life we *do* routinely make moral decisions by falling back on "intuitions"? When a jury decides that a defendant is guilty "beyond a reasonable doubt," for instance, there is no mathematical formula to dictate what counts as "reasonable." A hundred different factors may be weighed in the decision, and the conclusion is not the result of any formal reasoning process.

The crucial point is that these kinds of "intuitions" *result from* rather than *founding* the sum total of beliefs that result from our life experiences. In practice, we use "intuition" in the everyday sense as a shortcut method, just as we use our intuition to see that the diagonals of any rectangle must be equal, or to judge whether someone is getting ready to start a fist-fight. These "intuitions" are really just a cumulative result of our lifetime of history with other kinds of evidence and experience. We have informally measured many lines and angles during our lifetimes, and we have observed people's behavior when getting ready to start fist-fights. We also understand through phenomenological self-perception what our own behavior would be like if we were getting ready to start a fist-fight, as Edith Stein's work on empathy suggests. And we imagine what it would be like to measure the diagonal of a rectangle. The "intuition" is merely a summation of a lifetime of these other kinds of experience that we have gradually collected. All those other sources are the *real* evidence; the intuition that we form in real time is not the ultimate *reason* for the judgment. It is a pragmatically necessary short cut, based on earlier experience and thinking.

In a similar way, many experienced jazz musicians say they don't think about chords and scales and music theory when they improvise – they "just play" or "just feel what should happen next." Young musicians sometimes take this to mean there is no need to study chords, scales, and music theory – just "play what you feel." The effect, of course, is disastrous.

When we base everyday truth-seeking on an "intuition," we are not *starting* with an intuition and then "refining" it to make it logically consistent, as Moore and DePaul would have it. Quite the contrary, we are beginning with judgments that were based on other kinds of evidence, such as repeated past experience, or previous mathematical thinking, and *then* congealing the repeated experience into an abbreviated set of working assumptions about reality. The intuition is a *result* of all this repeated evidence, not the basis for it.

We are less seeking the truth than engaging in *casuistry* if we begin with value assumptions that literally have no more basis than "intuition," and then merely adjust the intuitions to render them logically consistent. We know in advance that we will end up with ethical viewpoints whose only recommendation is that they are not self-contradictory. The method of "refining" intuitions often becomes little more than a way to hold onto slightly tweaked basic beliefs that have no more grounding than habit, tradition, or psychological preferences. The slave-owners of the southern United States no doubt felt a strong "intuition" that slavery was a morally acceptable order of things.

Even though this "refining intuitions" approach seems obviously inadequate as a reason for beliefs, it could also be viewed as one of several primitive attempts to change from a foundationalist to a coherence epistemology. After all, a non–self-contradictory belief system *is* slightly more coherent than a self-contradictory one – *but only slightly more.* The coherence method needs more work.

In short, a workable coherence approach requires a "bigger tent." What I mean by "big tent" is this: In what I would call the *"smaller tent"* coherentist perspective (as technically laid out for example by Nicholas Rescher's *Coherence Theory of Truth*, 1973), a theory or viewpoint can never be completely justified, but some theories are more likely to be true if they satisfy certain criteria for coherence such as internal consistency, avoidance of proliferating ungrounded assumptions and ad hoc hypotheses, and a good fit to reasonably ascertainable empirical facts. But to use a "big tent" strategy requires expanding the criteria for coherence beyond just deductive logic and empirical evidence. Among other factors, it will need to include careful introspective or phenomenological methods as well as naturalistic investigation of the emotional processes that affect ethical thinking. The "facts" with which viewpoints must be reconciled include facts about human nature and the distorting effects of the specific motivations of the person or group forming the belief system.

An important caveat in the case of ethics is that we must also worry that theories advocating a *smaller number* of intrinsic values could end up enjoying an unfair advantage – that an overly *simplistic* theory could be judged as more coherent, just because it makes fewer assumptions. For example, contemporary policy analysts tend to work from a simple utilitarianism (see Ellis 1998), because utilitarianism requires only *one primary moral assumption* – that we should maximize the general happiness. Once this one assumption is made, it is only necessary to construct formulas to measure costs, risks, benefits, and harms. But does the simplicity and cleanness of such a process make utilitarianism the *correct* value system to use, especially when it conflicts with individual rights and fairness? This problem with overly simplistic forms of coherentism will be addressed in due course.

But first let's discuss a more pressing question about moral coherence: How does the hermeneutical process of worldview-formation actually work *in real life?* The role of "moral sentiments," as we have seen, should not be that they can by themselves ground ethical beliefs. But as already hinted earlier, moral sentiments do *affect our hermeneutical worldviews* by directly and indirectly shaping their motivation. Moral sentiments may also reflect informal coherence thinking that has previously occurred throughout the course of our lives.

6.2 The Hermeneutical Process: Why Religious Texts and Cultural Traditions Were Never the Real Basis of Anyone's Ethical Beliefs

The crisis of moral epistemology affects not only philosophers, but also philosophical laymen and the larger community. There was a time when many people believed that the truth about ethics was revealed through the authority of a religious leader or document. At least, many *believed* that they thought this; but then on careful reflection most people can come to realize (and increasingly so throughout history) that this was never the real epistemological basis of their beliefs in the first place.

Just consider any given issue on which different people disagree as to how a religious text should be interpreted. For example, the Books of Joshua and Samuel in the Bible occur just after Moses has revealed God's ethical commands not to steal or kill. But only a few pages later, the Hebrews reach the Promised Land, and we are told that it is already occupied by "a peaceful people."

But here is the interpretive problem: The scripture clearly states that God orders Joshua to commit genocide against this "peaceful people" in order to take their land. On almost every page of the Books of Joshua, Samuel, and several others, God is ordering the Israelites to take still another Canaanite, Baalite, or Philistine city by committing utterly unprovoked genocide against its people. If God is intelligent, we must assume that he wouldn't logically contradict himself by first telling people not to steal or kill, and then a few pages later explicitly ordering them to kill people in order to steal their land. So if we want to base beliefs on the authority of this document, we would first have to find a way to *interpret* the texts so as to choose between the contradictory passages and perhaps find a reason to reject or re-interpret one of them.

When we ask ourselves *why* we choose one item from a text as being preferable to another – for example, why we prefer "Thou shalt not kill" over "Kill everyone in the city" – the reason can't simply be that this statement is in the text. *Both* statements are in the text. So the reason for choosing one over the other must come from some informal philosophical thinking that occurs *outside* of the text itself, so as to justify interpreting the text in the way that we do. This extratextual philosophical process at least partially determines how we interpret the texts in the first place. Extratextual moral beliefs must *already* be in place to enable us to interpret the text the way we do.

The same problem arises even when there is no contradiction between different parts of a text. How do we interpret the statement in the Book of Isaiah that, if someone breaks any of a long list of minor rules about diet and sexual conduct, we should track down all of their descendants and cut their hearts out? The only people who would actually obey such a commandment are literally psychotics.

So the point is that what we believe from a text comes primarily from *outside* of the text. People throughout history are continually changing their religious beliefs to make them fit what seems to make more sense *from a philosophical perspective.* At the time of this writing, 44 percent of Americans have actually changed their religious denomination during their lifetime (see Pew Forum on Religion and Public Life 2015). And contrary to popular assumptions, religions don't just maintain the same dogmatic belief system from century to century. Religions evolve as their believers change their philosophical thinking. By the eighteenth century, the Catholic Church had for the most part discontinued the practice of executing opponents of the geocentric theory of astronomy; and in 1992, the

Church officially apologized for its condemnation of Galileo over this issue (*New York Times* 1992).

Even within the *same* religious denomination *at one given time*, people often disagree on the most important ethical issues – birth control, capital punishment, gay marriage, racial justice, etc. Both slavery and the abolition of slavery were defended by quoting the same religious book.

For most people, one of the main reasons for preferring "Thou shalt not kill" over a supposed command to commit genocide is that the former seems to *make more sense* from the standpoint of most people's philosophical thinking, whether they have had formal philosophical training or not. Granted, different people think about these philosophical issues differently. The point for now is that most people who believe in a religious text, even if they don't clearly notice this fact about themselves, evaluate and interpret their chosen text in light of philosophical thinking *that is not simply based on the text itself*. Many simply avoid reading the objectionable texts, but what motivates *this* selective inattention also comes from outside of the text.

And this means that there must be some philosophical beliefs that are *not* based on religion, and *on the basis of which* people interpret religious texts in one way or another, as well as selectively ignore some texts while highlighting others. This was what Schleiermacher (1838/1998) meant by "hermeneutics" in the theological context.

This epistemological principle has led to an increasing trend toward hermeneutic philosophy in recent years. Hermeneutics in the philosophical context extends the principle of textual interpretation to our interpretation of real life situations: *All* interpretations are filtered through philosophical presuppositions that don't arise merely from the situation itself. The way we see any situation is based partly on an already presupposed overall worldview, in terms of which we interpret the situation.

This overall worldview, according to the hermeneutic perspective, often acts as a "filter" through which we view reality, unaware of the prejudices and preconceptions that distort what we see. The psychiatrist Ludwig Binswanger (1963) used Heidegger's *Being and Time* as a basis for analyzing this situation in terms of an "existential *a priori*," which was actually quite similar to the way H. G. Gadamer formed his concept of "hermeneutics" (see Gadamer 1971). Our perception of reality is always "filtered" by these presupposed categories and preconceptions.

A classic real life example of this aspect of the hermeneutical process is revealed in a social psychology study of subjects who persisted in believing that Saddam Hussein was connected to the 9/11, 2001 attacks on the World Trade Centers and other US targets (Prasad et al. 2009). The interviews were conducted in 2007, long after it was well known that the alleged connection to Saddam Hussein had been factually debunked, and President Bush himself had made clear that he no longer believed there was any connection. The 49 subjects in the study were presented with the contrary evidence, including Bush's own admissions that there was no connection between Saddam and 9/11, yet *only one* of the 49 subjects was willing to give up the opinion that Saddam must have been involved in the attacks. It seemed that the idea that President Bush would take the country into an irrelevant war in Iraq based on completely false intelligence didn't fit into their *overall worldview.* When the facts don't fit our overall worldview (which is built up from a lifetime of experience and thinking), we either deny the facts, or find a way to modify our overall worldview while still retaining its general coherence.

And yet, as an example of the contrary tendency, much of the "cultural revolution" that occurred in the United States during the late 1960s was motivated by the need to re-evaluate established worldviews because of their incoherence with increasingly acknowledged facts about the Vietnam War and the widespread discrimination against Blacks and women in the workplace.

The Prasad et al. study suggests still another complicating factor for an "inner conflict" approach. There is an intricate subtlety in the way people regard "truth" claims in modern and postmodern cultures (if not in all cultures). The subjects in the study *knew* that President Bush had admitted that Saddam was *not* connected to 9/11; yet some of them said that the reason they still believed Saddam was connected was that they assumed that President Bush himself still believed it *even though he had stated otherwise.* When politicians and business or religious leaders make statements, they often don't intend their statements to be taken either as clearly true or as false. In many cases, what they want is that those who prefer to take the statement as true will do so; but at the same time, they also want others to assume that they are only pretending to believe what they say. They want people to cheerfully assume this disingenuousness and then interpret the statement as meaning something different from what is actually said.

A clear example is the "dog whistle" in political statements: The politician wants the larger *general* public to take at his word that he repudiates a given extremist viewpoint, while his more extreme followers assume that he *doesn't really mean* what he says when he repudiates the extreme view. When Donald Trump in 2016 publically retracted his famous false claim that President Obama was not born in the United States, polls showed that approximately half of his supporters assumed that he *still believed* Obama was not born in the United States, and was making the new statement merely for political reasons (Huffington Post 2016). The other half assumed that Trump *had always known* that Obama was really born in the United States, and was claiming otherwise just for rhetorical effect – that the false claim was merely a way of insulting Obama, as a teenage bully might accuse someone's mother of being a prostitute. That is how complicated the conflict can become between the truth-seeking and the confabulatory motives that are continually fighting against each other in our innate emotional nature. In 2016, many pundits even admonished the media for "taking Trump literally" (for example, see Zito 2016) – based on the premise that Trump's supporters knew he did *not* literally mean what he said. As a result, many pundits now speak of a "Post-truth America."

For our immediate purposes here the crucial point is that, when people say that they base ethical beliefs on religious texts, they can't *literally mean* what they say. They can't literally mean this, because they can't *interpret the text itself* literally. They must already have philosophical beliefs and attitudes that determine how they interpret the text, not to mention their *choice* of text itself.

6.3 A Similar Hermeneutical Process for "Cultural Traditions"

Many people speak as if they could hold moral views "because they were raised with certain beliefs" or "because they come from a certain cultural background." But it is easy to see that the same hermeneutical process we just observed with respect to religious beliefs also applies to "*cultural practices and traditions*" (as philosophers like Alisdair MacIntyre like to call them). The various members of a given culture are no more uniform and monolithic in their belief systems than are religion members; like religions, cultures change through history as a result of the thinking of their members. Anthropologists now realize that a specific archeological dig represents only a snapshot within a continually unfolding and changing evolution of the culture in question.

During the twentieth-century struggles for racial justice in South Africa and the United States, a choice had to be made between a dominant traditional interpretation of existing cultural mores and a new alternative – and this choice itself had to be made independently of the tradition itself, for the same reason we just discussed in relation to religious texts. We decide which texts to accept and how to interpret them based on extra-textual considerations. Similarly, whether we accept or reject a tradition, or what the tradition even means, is determined by philosophical assumptions that we bring with us to our interpretation of the meaning of the tradition.

MacIntyre (1984) tries to maintain that the reinterpretation of a tradition comes *from within the tradition itself*, not from outside of it. For example, he points out that the US democratic tradition includes not only majoritarian elements, but also guarantees of individual rights against an overbearing majority. But what MacIntyre downplays is that whether we choose to prioritize one or the other of these aspects of the tradition must come from outside of the tradition. *Both* elements are in the tradition, so the decision as to which should be prioritized, if not purely arbitrary, requires some additional thought process with which to make such a decision. There are also aspects of the tradition itself that we may choose to reject altogether – for example, the explicit endorsement of slavery in the original US Constitution.

Even if MacIntyre maintains that the principle that "slavery is wrong" was already implicit in the original document, the endorsement of slavery was *explicit;* the document itself and the tradition flowing from it do not provide any decisive basis on which to decide that one element or the other was to be extrapolated at the expense of the other. What we accept, reject, or prioritize in a tradition ultimately comes from some thought process external to the tradition itself. In the same way that the choice between "Thou shalt not kill" and "Kill everyone in the city" requires assumptions from outside of the texts, so the choice between slavery and universal political rights requires similar assumptions that must come from some other source than a mere "cultural tradition."

This brings us back to our original question about the epistemological grounding of ethical beliefs. Throughout the modern era, people have increasingly recognized that ethical beliefs can't simply be grounded in the authority of a cultural tradition any more than a presumably literal interpretation of a religious text, and for the same reasons. The way we choose, interpret, and respond to traditions must result from our philosophical thinking, for better or worse.

But at the same time – and for good reason – people are also mistrustful of the ability of any *philosophical* method to demonstrate the truth of ethical propositions in a purely logical or rational way. Worse still, as recent postmodernists have been at pains to point out (for example, Foucault 1994, Lyotard 1984), our own hermeneutical maneuvers in principle tend to remain invisible *to us.* It is difficult if not impossible to know what our own distorting filters are blocking out, because the very fact that they distort reality is *itself* blocked out.

So on the one hand, people are increasingly coming to realize that philosophical assumptions shape the interpretation of religious texts and sociopolitical traditions rather than the other way around. But at the same time, we also have been increasingly up in the air as to whether philosophical thinking can provide reliable epistemological criteria for ethical beliefs at all. The two most obvious epistemological criteria that might claim to ground the truth of statements are *logic* and *empirical science.* But many philosophers, from G. E. Moore to A. J. Ayer to the mainstream of the twenty-first century, now tend to agree that it is impossible to prove any significant ethical principle either by means of empirical evidence *or* through deductive logic, or even a combination of the two.

6.4 The Problem of Moral Epistemology as It Affects the Hermeneutical Process

As a result of these hermeneutical demands, it is all the more necessary to confront head-on our second reason why ethics is a "dismal science": There seem to be no reliable methods of demonstrating the truth or falsity of potentially competing moral viewpoints. Given the minefield of opportunities for hermeneutical self-deception combined with this unfortunate *genuine difficulty* of philosophical value theory, we must finally ask ourselves whether it is even appropriate to deal with these problems by means of a "foundationalist" theory of knowledge, in which we attempt to prove the truth of statements using unquestionable epistemological assumptions.

But "dismal" doesn't entail "hopeless." Even if we admit to ourselves that there is no such thing as absolute proof when it comes to ethical viewpoints, this should not mean that moral skepticism automatically becomes the default position in case of uncertainty. As Robert Almeder (1986) argues with respect to scientific theories, it may be that the best we can do is to choose what seems to be the most coherent of a number of viewpoints, admitting that the process will always be fallible, but not arbitrary or completely groundless. And of course extremely skeptical viewpoints would be among those to be considered.

"Coherence" here might be defined in terms of criteria that increase the likelihood of a truthful outcome. Ethical egoism, relativism, and nihilism would be among the theories to be included in comparisons of the coherence of various positions. Extremely skeptical viewpoints would have no special status as the automatic outcomes in case nothing can be absolutely proven. If we demand completely certain foundationalist proofs, it is all too easy for the moral skeptic simply to sit back and demand, "OK, prove to me that XYZ is wrong; if not, then XYZ can be assumed to be just as justifiable as any other course of action."

Even those who don't consider themselves skeptics can easily reject any principle they choose, as justified by the *failure to prove* the validity of that principle – thus playing into the confabulatory tendency to believe whatever we please. Moreover, even if philosophers could establish some correct theory by means of logic and/or empirical evidence, we would then be expecting everyday people to have PhDs in philosophy in order to be able to discern the truthfulness of the theory.

But consider the analogous case in non-moral contexts: When ordinary people pay appropriate attention to the relevant considerations, they can easily see that the idea of global warming is more coherent than the denial of global warming. To deny its reality requires too many ad hoc hypotheses to explain away conflicting information. So why can't similar coherence criteria be used for comparing different moral viewpoints? Of course, as in the global warming case, this would require a motivation to pay appropriate attention to the relevant considerations.

The central strategy of the coherence approach is that ethical theory should be structured in a way more like science than math. There are no unquestionable initial axioms, nor uniquely deducible theorems. But there are facts in the constitution of human emotion, including not only emotions that warp our moral viewpoint, but also the sheer fact that humans can't take the position that nothing has any value; nor can we coherently ignore the fact that other valuing creatures "afford" valuing to the extent that we pay attention to their valuing dimension – which leads us to think about their sense (3) value. Without value commitments, we would be unable to act. Beyond that, we can try to construe values as limited to our immediate family or tribe, or universalize them; we can try to construe them in terms of utilitarianism, retributive or distributive justice, emotivism, egoistic hedonism, or some other relatively coherent ethical viewpoint.

At that point, we are already involved in the coherence project of trying to hold together a viewpoint that doesn't depend on fallacious reasoning, nor contradict itself or our best honest assessment of the facts of the emotional lives of valuing creatures, including ourselves as valuing creatures. We would also of course need to avoid the naturalistic fallacy, and to eschew reliance on "moral intuitions" and proliferation of ad hoc hypotheses. And we would need to devise ways to understand how our specific emotions may be warping our hermeneutical interpretation of moral reality.

Granted, unlike in scientific work, the preconceptions in ethical thinking that motivate fixing new facts to previous theoretical commitments are much more influenced by inner emotional conflicts. These conflicts pose an additional hurdle when it comes to ethical thinking.

Inevitably, a coherence approach must make a good bit of room for careful clarification of our attempts to introspect into the emotional processes that complicate ethical thinking and thus influence our overall hermeneutical worldview. The sheer fact of the hermeneutical process means that justifications for behavior may be motivated by the need to maintain an entire worldview that has been formed for emotional reasons over a lifetime, or even over the course of many generations or entire epochs of history. Some of the relevant affective processes may be inevitable features of sophisticated conscious creatures per se – not merely accidental products of history.

The unreliability of moral emotions interacting with everyday hermeneutical gymnastics might not be so discouraging to the ethical theorist if it were possible to ground ethical statements in rational proofs or logical inferences from non-controversial empirical facts. But both academic philosophers and the general culture have lost confidence in the ability of logic and/or empirical evidence to yield moral conclusions, unless we have already assumed moral premises. And this would mean that those premises themselves remain ungrounded and irresolvably controversial.[2]

I am by no means suggesting that logic is not an important tool in evaluating moral viewpoints. So is empirical evidence, especially empirical evidence about human nature and the social sciences. But if logic and empirical evidence are the only tools in the tool box, we still haven't narrowed the range of possible viewpoints that pass the test of coherence about equally well, or even gotten them all into the same ball park – for instance, the utilitarian ball park or the "moral desert" ball park or the "justice as fairness" ball park, or for that matter the many different varieties of distributive justice theories that can be argued. We are going to have to add more tools to the coherence toolbox – including not only the hermeneutical effects of specific emotional dynamics, but also the effects of general facts about human affectivity *on moral epistemology* in general. The next section will focus on why this last point is so crucial.

6.5 The Role of Complex Emotional Histories in Shaping Moral Ambiguities

When people make statements like "We always ought to maximize the general happiness" or "Individual rights and distributive justice sometimes ought to supersede the value of the general happiness," we have to unpack what various speakers mean by the claims they are making. Even after we acknowledge that a given word can have different meanings for different users in different contexts, the problem remains that the clarification of any one of these meanings not only requires phenomenological reflection into the speaker's own meaning intentions, but those intentions in turn may also *presuppose the truth* of still other assumptions. Presuppositions are already nested within concepts.

For instance, if we question someone's statement that some criminal "deserves" a harsh punishment, the response often will be that "chaos would result" if criminals go unpunished. But then if we ask whether this means that criminals ought to be punished even when doing so does *not* contribute to deterring crime, or perhaps even if it causes an increase in crime rates by converting minor offenders into hardened career criminals, the response often is that "desert" means more than just an instrumental value intended to deter crime – which apparently was what was intended in the initial response. Already, the lack of clarity as to what is intended by "desert" is becoming obvious (Sher 1987 discusses the variety and typical ambiguity of such meanings). Further clarification will show that the speaker's intent about "desert" *presupposes certain alleged facts* about social conditions, the free will–determinism issue, etc. For example, those who have actually witnessed executions are overwhelmingly more opposed to the practice than those who have not (Dieter 1995).

And the problem remains, even after we do clarify what someone intends by "desert," why anyone should believe that punishing according to "desert" in that particular sense has intrinsic value rather than only *instrumental* value when and if it serves as a means to an end (for example, deterring crime and/or teaching the offender a lesson). By what criteria do we decide that, say, a bank robber "deserves" 20 years in prison rather than 10 or 30?

Criminologists who study "common intuitions" about what punishments are "deserved" (for example, Sellin and Wolfgang 1964; Roth 1978) reveal a great deal of confusion, variability, and inconsistency. Roth finds that the "desert" judgments of prosecutors, defense attorneys, policemen, and others working within the criminal justice system – regardless of whether they work on the prosecutorial or defense side – differ systematically from those of ordinary civilians. Because of the murkiness as to what we mean by such moral concepts and statements, a coherent epistemology needs to include some careful thinking about what kinds of *introspective* or *phenomenological* methods are most helpful in clarifying what various people might mean by claims about "oughts," "deserts," or "rights" – let alone why we should believe that certain statements using these concepts are true or false. Here again, the hermeneutical process intervenes between moral concepts and our own introspective access even to what we ourselves *mean* by the concepts.

Even if the answers to none of the important questions in this domain can be proven by means of an absolutely well-grounded argument free from controversial first premises, emotivism is still not our only remaining option. Granted, the role of moral emotions must become more prominent in light of the murkiness of the meaning of these concepts. But we can avoid a simple emotivism if we shift gears from the traditional "foundationalist" epistemologies; we can acknowledge that ethical statements are not strictly provable, but that there are criteria for the evaluation of the coherence of an overall philosophical viewpoint. Developing a workable coherence strategy, however, is not as easy as one might assume.

In addition to the obvious hermeneutical process in which we weigh new experiences in relation to an already established moral worldview, a further aggravation of this limitation is the legitimate difficulty of ethical epistemology. This genuine difficulty may then play into the hermeneutical distortions of our moral perspective, without our having any way to determine whether or to what extent this is in fact what is going on – again, because of the hermeneutic circle. The next section explores this additional complication that stems simply from the failure of logic and empirical evidence to be able to "prove" that one moral viewpoint is better than another.

6.6 Both Deductive Logic and Empirical Evidence are Inadequate for Establishing the Coherence of Viewpoints

We have come almost full circle from Descartes' dream of logically deducing all important philosophical conclusions from one or a few absolutely unquestionable first premises. By mid-twentieth century, the young A. C. Danto (1968) was in the vanguard of a newly emerging consensus that deductive inference is almost always purely "garbage in, garbage out." Ethical statements, in particular, are not the kind of statements that can be deduced through sheer logic without first assuming premises that already are subject to controversy, and also are already laden with ethical content. The reason is that a deductive argument in support of a substantive conclusion requires substantive premises. Given the premises

P implies Q
Q implies R,

we can prove that P implies R. But, obviously, the question of the truth or falsity of the premises themselves must ultimately depend on some other kind of evidence than deductive logic.[3]

Moreover, even if we could grant non-controversial assumptions about the way the world is – for example, that humans value happiness – this assumed "*is*" by itself still can't imply "*ought.*"

For instance, suppose we begin with the non-controversial premise that most drivers exceed the speed limit at some time or other. And suppose we then were to try to make the following inference:

Most people exceed the speed limit.
Therefore, people ought to be allowed to exceed the speed limit.

To move from the premise to the conclusion, we would already need to have assumed the truth of an "ought" statement – for example, "people ought to be allowed to do things that most people do." In effect, the argument is tacitly assuming the very principle that it superficially appears to be proving. So it seems impossible that anyone could deduce an "ought" conclusion without already assuming an "ought" premise.[4]

Someone might think we could simply *define* "ought" to be equivalent with some "is," as for example Euthyphro in Plato's dialogue tries to define what we ought to do as equivalent with "whatever the gods command." But then the argument becomes an empty tautology and has no substantive content. There would be no way to question whether we ought to do what the gods command if the word "ought" has been arbitrarily *defined* to automatically entail this conclusion.

In light of all these epistemological difficulties, the next chapter will explore the possibility of eschewing foundationalist "proofs," and instead using a coherence strategy to navigate the hermeneutical situation in such a way as to lead to reasonable choices about what is more or less likely to be true in ethics, at least as best we can honestly assess it given the facts and philosophical ideas available to us at the time within the limits of our imagination and creativity.

We can begin by looking at some historical examples of the way "rational proof" has evolved into more of a coherence strategy. Perhaps surprisingly to some, part of the reason for this evolution is the way traditional ethical theorists like Kant, Gewirth, and Rawls have attempted to understand the interactions between reason and emotion when we engage in moral thought.

NOTES

1 I use "moral intuition" not in the more general sense, but specifically as applied to ethics. Even self-consistent "moral intuitions" can contradict each other; but it should be *impossible in principle* for well-founded statements to contradict each other. So any grounds for choosing to accept a particular intuition must rest elsewhere than the intuition itself. Not just some, but *all* intuitions lose their justificatory force for this reason.

2 Can a purely logical argument tell us whether economic property rights are more or less important than the ideals of equality and fraternity and the dignity of working people? John Rawls (1971/1999) gives us a rough formula for "distributive justice," but leaves out any argument about the relative priority of "moral desert" and other proposed values. And the "distributive justice" formula itself is open to questions and in need of assumptions, of which Rawls is well aware (see Rawls 1993).

3 Granted, some statements can be proven logically without premises – e.g., "all equilateral triangles are geometrical figures." But such statements are empty tautologies; no ethical statement with real content can fit into this category. If we say "Democracy is majority rule, therefore a country that has majority rule is democratic even if it denies minority rights," we merely announce that we intend to use one definition by preference over others.

4 This principle stands up even in the face of Kant's "Ought implies can." Granted, if people *can't* do X (in the relevant sense of "can't"), we can deduce that it is *not* true that they *ought* to do X. But it doesn't follow that people ought to do Y or Z or any other behavior.

CHAPTER 7

Kantian Abstractions and the Embarrassment of Reason

The Need for Hermeneutics

Immanuel Kant offers an unusually rich phenomenology of the "moral conscience" – that which is completely lacking in the case of the true sociopath. Kant's observations on this score are more subtle than is often appreciated, and they play a large role in his overall moral viewpoint.

Interconnected, however, is the little-noticed elephant in Kant's room: With regard to empirical psychology, he is dogged by the same extreme hedonistic assumptions that are still prevalent to this day. In the opening sentence of Theorem II of the *Critique of Practical Reason*, Kant writes, "All material principles – as such – are, one and all, of one and the same kind and belong under the general principle of self-love or one's own happiness" (Kant 1788/2002, p. 33). Or again, "[A]ll determining bases of the will except for the single pure practical law of reason (the moral law) are one and all empirical and hence, as such, belong to the principle of happiness" (1788/2002, p. 119).

Ironically, those same exclusively hedonistic psychological assumptions required his otherwise unnecessary excursions into the metaphysical realm in order to make morality possible – given that all non-metaphysical or "empirical" motivations were presumed to be oriented merely toward one's own happiness. How would Kant's approach have been different if he had been aware of the non-hedonistic *natural* emotional drives we have been discussing here?

Kant makes astute connections between the pangs of conscience and the inescapable demands of reason, which he observes we can't bring ourselves to ignore even though it would be more pleasant to do so. It makes some sense that, if we *can't will ourselves* to believe something we know isn't true, this fact could connect with the involuntary nature of the conscience. But a major drawback to this approach is that, in his handling of the somewhat involuntary "demands of reason," Kant lays the groundwork for a paradigm that still permeates much of moral philosophy – a reliance on the power of deductive logic to solve problems that can't be addressed empirically. We began to see the limitations of this method in our last chapter. Yet many philosophers, like Kant, assume that the role of a priori logical methods in establishing ethical beliefs can be so plain that ordinary people feel "pangs of conscience" when they violate the resulting principles.

Granted, ordinary people do use logic quite proficiently, as evidenced by the computer systems and automobiles we are able to produce and maintain. Kant's problem is not an over-estimation of people's ability to use logic. The problem is with the kind of logic Kant wants people to be limited to – ultimately, deductive logic. One of the consequences is that such purely a priori methods tend to remain at such an abstract level that what they mean for *real-life decision making* can become almost infinitely malleable.

As we have already seen, easy malleability in the application of abstract principles can be deadly. It quickly becomes one of the seductive temptations inherent in the potentially self-deceptive aspect of the hermeneutical process. I will show later that, even if Kant's theory of the "categorical imperative" were accepted as the basic principle of morality, its implications for what should be done in specific situations would remain almost completely undeterminable.

But on the positive side, much can be learned from theories like Kant's about the psychology of the "moral conscience." In the final analysis, philosophers like Kant, Gewirth, and Rawls are beginning to move us toward a coherence rather than foundationalist epistemology for ethics. The essential problem is that we need to expand the toolbox beyond merely deductive logical inference.

7.1 Kant and the "Humiliation" in the Face of Reason

We already saw why ethical statements can't be proven with deductive logic – even if we begin with minimally controversial non-moral assumptions. But we also saw that ethical statements ultimately *depend* on philosophical reasoning. Neither mere cultural tradition, nor altruistic and cooperative feelings, nor religious texts can determine what we believe. Ultimately, the way we use those sources depends crucially on our philosophical reasoning – even if only implicitly. Absent a cogent philosophical approach, we would remain stuck in the no man's land of emotivism and "anything goes" cultural relativism.

But if philosophical thinking is our guide, then the most powerful tool of philosophy does seem to be deductive logic. So there has been no shortage of philosophical attempts to meet the daunting challenge that our previous chapter characterized. We simply can't leave ourselves permanently stuck in the untenable position of having no moral principles that we can really believe in. The alternatives seem to be mere social conventions or arbitrary emotional preferences.

The situation is complicated further when philosophers pay too little attention to the relationship between reason and emotions, as if the two could be starkly separated. When we think in terms of a "battle between the head and the heart" (as Thomas Jefferson put it in one of his diaries), we tend either to use logic as the *only* tool, or to relegate moral principles to "mere emotional preferences."

Kant, to his credit, avoids this trap. Rather than rigidly opposing emotion against reason, Kant's idea is that "respect for the moral law" – an emotion in its own right – is sharply distinguished, not from emotions generally, but rather from all the *"empirical"* emotions.

The "empirical" emotions are equivalent with what we now would call "natural" emotions. Kant calls them "empirical" emotions both because (in Kant's view) they take *empirical* affairs as their objects, and also because "nature" is defined as that which in principle could be an object for *empirical study.* In this sense, the supposedly *non-natural* (supernatural, metaphysical) feeling of "respect" is not only distinguished from anything that is empirical (natural), but Kant thinks it is also outside of the chain of causes and effects. If it were in that chain, it would be natural, "empirical," and thus reducible (or so Kant argues) to the desire for happiness or pleasure (Kant 1788/2002, Theorem II).

Notice that Kant's view of the "empirical" emotions *is as hedonistic and reductionistic as that of any twentieth-century behaviorist.* They all reduce to a desire to maximize pleasure and minimize pain. This is precisely why, in Kant's analysis, they can't ground any moral principles in any non-trivial sense.

To many modern scientific ears, Kant's idea of a non-naturalistic or metaphysical feeling of "respect for the moral law" often sounds quaint or even non-sensical. But only a slight modification of Kant's affective scheme is needed to make it fit into the contemporary array of innate emotional systems. Kant's mistake is to assume, like the twentieth-century behaviorists criticized by White and Panksepp, that all natural ("empirical") emotions must be motivated by a desire to achieve *hedonistic* aims, in the sense of consummatory, instrumental reward for behaviors. Contemporary emotion research, as we have seen, shows that this is *not* the case.

Thus we can now say that "respect," which is driven by an innate exploratory drive (Hume's "love of truth," which Kant seems to have missed in his reading of Hume), is a *natural* part of the emotional equipment of humans. We desire to know the truth about ethics just as about other things, independently of whether a particular truth makes us happy. The individual organism is wired to engage in this type of activity without any reward that is external to the behavior itself.

If Kant had allowed that some *natural* emotions drive the search for truth and are not in the service of the pleasure/pain system, then he could have allowed that "respect for the moral law" is a natural emotion (i.e., in Kant's terminology, an "empirical" emotion) – a result of Panksepp's SEEKING system, and more specifically the exploratory aspect of the SEEKING system.

An additional benefit would have been that, if the drive to use reason had been treated as an independent *natural* motive, then it would not have been necessary for Kant to posit that humans have a "contra-causal free will." The kind of freedom required by realist ethics is that people can make choices that *determine their actions* – unimpeded by forces that are currently external to them – *regardless of whether they have been predetermined* to be the kind of person who will choose X over Y. This kind of freedom doesn't require metaphysical commitments. In terms of modern self-organization theory, for instance, we could say that people are systems that appropriate, reorganize, and replace their own micro-constituents as opposed to being the mere passive result of inputs from the environment. In this case, their actions would be controlled (at least partly) by their choices.

Whether or not an efficient causal chain had led to the *etiology* of such a self-organizing system (i.e., had caused it to develop the way it did), we would still be self-organizing in our real-time activity – not merely *re*-active to *currently* external inputs – and thus we would still be capable of making choices. Even if the past, P, had completely determined my character, C, which in turn determines a future action A, the fact would remain that C determines A, and it would do so by virtue of the choices I make. Whether or not I was caused (by genetics and environment) to have the character of someone who would choose to diligently practice a musical instrument, the fact remains that the reason I practiced was because I chose to; no currently external force caused or forced me to practice. So whether determinism is true or not, we have already seen that people can be motivated to respect what they take to be a truth, just as Kant wanted to argue. The feeling of "respect" would not have needed to stand out from the natural realm of causal processes.

Nor would Kant then have needed to create irresolvable contradictions between human nature as available to empirical social sciences, and on the other hand the human soul in its "noumenal" or "metaphysical" being, which supposedly would have to stand outside of the temporal realm of causes and effects, and remain inaccessible to empirical and phenomenological inquiry. To a great extent, by trying to *exile* the empirical social sciences from ethics, he invited them to counterattack and ultimately to *take over the entire kingdom* when people finally became skeptical of Kant's metaphysical speculations. And since the social sciences at that time were purely hedonistic, they ended by destroying the kingdom, which ironically was Kant's worst fear.

Despite this unnecessary metaphysical turn, a more naturalistic construal of the emotion of "respect" could make Kant's approach compatible with contemporary neuropsychology. Given this modification, Kant's psychology is based on a more sophisticated phenomenology of the emotions than he is often credited for – an analysis of the *involuntary* nature of the "pangs of conscience." If this insight can be updated, it might provide some groundwork for why the "love of truth" could also motivate an emotional "respect" for the truth once we feel that we have some idea of what *is* true.

If we do modify Kant's view of the "natural" emotions to reflect the new non-behaviorist neuroscience, we end up with an analysis that for the most part can be very helpful in the current crisis of moral theory and moral psychology. So let's take a closer look at Kant's emotion of "respect."

(a) Respect, "humiliation," and the pangs of conscience. Kant's assessment of the idea of "respect" is detailed enough that he actually offers a fairly good phenomenology of the experience of the "moral conscience." As noted earlier, the contemporary psychiatrist Martha Stout (2005), who incidentally cites Kant favorably on this score, summarizes research showing that 4 percent of Americans are true sociopaths – meaning, by her definition, that they literally never experience any pangs of conscience. This means that at least 96 percent of us *do* suffer from having a conscience. What is this anti-hedonistic cloud that persistently hovers over almost all of us?

Kant is correct to point out that pangs of conscience are experienced as *involuntary*, since we would prefer to avoid such unpleasant feelings if we could (Kant 1788/2002, pp. 98–103). "Respect is a tribute that, whether we want to or not, we cannot refuse [to pay] to merit; we may perhaps hold it back outwardly, yet we cannot help feeling it inwardly" (p. 101). Kant notes that we may have divergent and sundry views of whether a given action *actually is* harming others in ways that are against the moral law. But if we do *believe* that we are harming them in this way, we can't escape the pangs of conscience.

Kant thus says that conscience prompts a feeling of "humiliation," in a dual sense – first, in the sense of rendering us "humble" (pp. 98–103). But Kant also means that, if we try to withhold respect from that which we *must* respect by virtue of our rational nature, we are "humiliated" in the stronger sense of the word (p. 98 and passim). According to Kant, the reason for this involuntary feeling of humiliation is that an intelligent being can't avoid acknowledging the moral truth, because of the rational self-evidence of that truth. Nor can we avoid respecting it once we have seen it, because our basic nature is to seek the truth independently of pleasure/pain motives (see also Kant 1797/1996). We feel "humiliated" by the fact that our actions fail to accord with the moral truth, yet we can't turn our backs on the truth.

So the *painful* nature of the pangs of conscience can't cause us to turn away from them, and thus can't deter us from submitting to their tyranny. In terms of contemporary affective neuroscience and psychology, we can view this non-hedonic love of truth as a *natural* motive. The love of truth is rooted in the exploratory drive, and in Panksepp's SEEKING system.

Also to Kant's credit, he avoids the naturalistic fallacy by acknowledging the role of conscience, *not* as a mode of "moral intuition" (which Kant 1788, p. 57, *explicitly repudiates*), but only as a driving *motivation* for counter-egoistic respect for the moral law – despite and even because of the fact that it "humiliates" us. Kant is aware that the conscience can't be used as a *method of knowing* what is right or wrong, because whether the conscience bothers us or not is largely determined by whether we already believe what we have done to be wrong. It follows that whether we believe it to be wrong can't be *determined by* the resulting pangs of conscience.

When our beliefs begin to be challenged as incoherent, however, the feelings of "humiliation," and thus in effect the "pangs of conscience," do begin to emerge, because we begin to suspect that we have disrespected a moral truth. So we can see that the *negative* feeling of pangs of conscience is inextricably intertwined with the *positive* feeling of respect for the moral truth, or wanting to know the moral truth – an aspect of the exploratory drive combined with other overtones of the more "social" emotions.

Kant is aware, of course, that this peculiar feeling of humiliation usually occurs in a mixture with other feelings – FEAR of being caught, PANIC over the possible resulting social isolation, and even PAIN resulting from the anticipation of punishment. But these feelings can occur without a moral conscience. It is the "humiliation" aspect that uniquely characterizes the conscience; the involuntary nature of this feeling, in Kant's view, results from our inability to disrespect the truth, though we may try. Respect for truth on the one hand, and on the other hand humiliation by our pathetic attempts to ignore the truth are opposite sides of the same coin. To ignore the truth requires shutdown of the exploratory drive, which dampens the SEEKING system; this in turn severely diminishes our feeling of inspiration – our feeling that our lives are worth living, or that anything is worth doing. Guilt feelings share that aspect in common with depression, although they are different in other respects, especially in that they hyperactivate the FEAR and PANIC systems even while damping down the SEEKING system. But again, mere fears of punishment and/or social isolation don't add up to pangs of conscience without the additional component of *believing* we know what we have done to be really wrong, not merely judged unfavorably by others.

But things become more difficult when Kant tries to account for *how* we can come to know the truth. While granting the need for our understanding of the moral law to be coherent (rather than foundational), Kant wants to show that all moral theories other than his "categorical imperative" are logically incoherent. A quick recollection of Kant's argument reminds us that this approach, while correct in taking essentially a coherence rather than foundational approach (because "practical" reason is not *pure* reason), he nonetheless oversimplifies the task at hand, and tries to approach it with only one tool – logic.

(b) How do we know this "moral truth" which humiliates us? One of the lessons that must be taken from Kant is that, despite his valiant efforts, we can't know the "moral truth" by deductive logic alone. To be sure, Kant does admit that we can't *logically prove* the truth, but he does think logic alone can shoulder the task of allowing a comparison of the viability of competing moral viewpoints. The problem is that, contrary to Kant's hope, *not all viewpoints except for Kant's* end up logically contradicting themselves. And this means that we need more tools than just deductive logic if we are to develop a coherence-epistemology tool box.

Some people used to believe that Kant's categorical imperative was logically deducible from the harmless minimal assumption that we need to adopt *some* ethical principle, and if so, then we want it to be non–self-contradictory. Such a "categorical imperative" asserts merely that we should not act toward others in a way that we could not be willing for them to act toward us. To some, it would seem self-contradictory to violate this principle. How can I say that it is wrong for others to steal from me without admitting that it is equally wrong for me to steal from them?

But Kant himself never intended his argument as a foundationalist proof, and for good reason. Instead, he is contrasting the coherence of his approach with competing viewpoints (such as utilitarianism) by means of the "categorical" versus "hypothetical" distinction: He first defines a "categorical" imperative to mean an imperative that is so basic that it is *not derivative* from some more basic imperative. "Hypothetical" imperatives *are* derivative from more basic imperatives (hence the term "hypothetical"). For example, in the statement "You should help others if you want to be truly happy," the "ought" claim, "You should help others," is dependent on the more basic *hypothetical* premise – "if you want to be happy." But if we ask why people want to be happy, the answer is a statement of psychological *fact* rather than an *ethical* imperative. So hypothetical imperatives (such as utilitarianism) are not really ethical principles at all. They are only prudential considerations that serve *psychological preferences.* (Kant's clearest exposition of this point is in Kant 1788.)

In this regard, Kant actually anticipates G. E. Moore's idea of the "naturalistic fallacy." There is no point in saying that people should do X because they naturally want to do X. That would reduce ethics to a "hypothetical imperative," which would be devoid of any non-trivial ethical content.

Kant concludes that only a "categorical" imperative can be a genuinely *ethical* principle (as opposed to a means of serving a psychological preference). And the only categorical imperative that can be logically justified is "Do not act according to any maxim that you could not be willing for others to act by."

Although this principle is not derivative from a psychological preference, it seems *psychologically undeniable,* because we can't deny it without willing something that we *can't will* – namely, that others should be allowed to treat us harmfully and without regard to our well-being. Therefore, we would be logically contradicting ourselves if we were to claim that it is all right for us to treat others in those ways. The logic of the principle is not based on the sheer fact that we will this or that, but rather that we can't will something and will its logical contradiction at the same time.

Kant thinks we can't escape this moral law without rendering our reasoning incoherent – that is, contradicting ourselves. And psychologically, as rational beings whose nature is to seek the truth, we can't avoid *respecting* these dictates of reason. When we try to escape them, we end up feeling "humiliated" in the form of pangs of conscience.

Why, then, are there still moral controversies in the twenty-first century? We must take somewhat seriously the reasons why things aren't as simple as Kant wants to make them.

(c) Why logic alone can't establish the "moral law." A couple of centuries of Kant scholarship have suggested many problems with such an unambiguous determination of the objective moral law through reason alone. In the first place, Kant's categorical imperative is *indefinitely malleable.* I can, without logical self-contradiction, will that no one should help the poor, even granting the risk that I myself might someday be poor. I can be willing for even *stealing* to be legal if I also can be willing to live in a world where everyone (including myself) is individually responsible to protect *themselves* from thieves – for example, by using their guns. Obviously, I *can* will these things if I feel that I personally can protect myself from such adversities, or even if I happen to be a daring risk-taker. Someone will say: Yes, but it would be foolish to think we don't need protection from thieves. Foolish, perhaps; but not logically self-contradictory.

Besides the lack of self-contradiction, it is also unclear how to determine what specific standards of conduct would be required by the categorical imperative. A slave owner can will *without self-contradiction* that anyone who is born a slave should be a slave, while cheerfully acknowledging that the same principle would apply to himself if he were born a slave. This might be a foolish thing to cheerfully acknowledge, but it is not self-contradictory.[1]

Someone may say that this interpretation misses the point. The one thing I *cannot* consistently will is that others treat me "as only a means and not at all as an end." But the problem here is that, if I simply give my slave a small Christmas present, or an occasional fifteen minute water break, I am not treating him *only* as a means. And of course we *all* treat others *partly* as means, for example when we make contracts with them or benefit from their good work. Obviously, when I compete with other candidates for a job, I am treating them at least partly as means rather than ends. But the requirement to treat people "partly" as ends is in principle an *indefinitely low bar* – as in treating a slave as an end by giving him a fifteen-minute water break. So here again, the principle as Kant formulates it is indefinitely malleable.

Moreover, the obligation to obey the categorical imperative itself is not logically provable. *Why* should I not act according to a principle that I couldn't will to be a universal guide to human conduct? I can't will that everyone in the world become a saxophonist (because then there would be no plumbers, electricians, or clarinetists); but does this entail that I should not become a saxophonist? And if not, then exactly what *does* it entail?

It is not logically self-contradictory for me to become a saxophonist and yet will that others not do so. Similarly, there is no logical inconsistency in saying that I don't want others to steal from me, but that I do want to steal from them. Perhaps it is morally wrong to steal, but this argument doesn't prove that it is.

Kant himself was aware that we can't deduce "stealing is wrong" from "I cannot will a situation in which others are allowed to steal from me." This is why, in the *Critique of Pure Reason*, he carefully separated out questions that human reason by its very nature is ill-equipped to resolve theoretically (such as "Is the universe finite or infinite?") – and he specifically included ethical questions in this category.[2]

Kant believed that we first had to adopt the premise that, for *practical* purposes, we can't understand the world as a world where there are no moral principles. We can't understand such a world because we already find ourselves to be guided by moral principles. Kant often calls such guidance by moral principles "freedom"; he takes the fact that we need such principles to guide us as simply a fact of life. Therefore, we can't make sense of the meaning of our lives without some such principles. Once this has been assumed, then Kant thought we must adopt non–self-contradictory ethical principles, since we have decided on practical grounds that we must adopt *some* ethical principle. But a "*hypothetical*" imperative, as we have already seen, *is not a genuinely ethical principle;* so we need a categorical one. His point is that it would be logically inconsistent to adopt a genuinely ethical principle of conduct for others that we could not be willing to follow ourselves.

The crucial problem is that Kant thinks logic alone can provide sufficient grounds for eliminating all competing approaches other than the categorical imperative. In effect, he is using a coherence approach (by making ethics a matter for *practical* reason), but with only one tool in the tool box – deductive logic. Even if it *is* assumed that we need to believe in some non–self-contradictory ethical norms, it still is not clear that self-contradictions would arise in the way Kant thought in all of the competing non-Kantian viewpoints (for example, utilitarianism, Rawlsian distributive justice theory, or some other moral or social theory). The reason slavery is wrong is not because it would be *logically self-contradictory* to say that anyone who is born a slave should remain a slave.

A rich man can assert, without logical self-contradiction, that if he were poor he would be willing for the rich not to help him. He can assert without self-contradiction that the rich should pay no taxes, or that it should be permissible for political positions to be bought by the highest bidder, that political leadership can be hereditary, or that legislators can be bribed. Food manufacturers can assert without self-contradiction that use of harmful food additives should not be regulated, and that even if they themselves were not food manufacturers, they would still be willing to live according to this principle. There is no logical self-contradiction in a mafioso's assertion that anyone who feels they can get away with killing and racketeering should be free to do so, and that people simply must assume their own responsibility for protecting themselves when others try to do the same things to them.[3]

What we decide we would be unwilling for others to do to us (or try to do, if they can get away with it) *varies* – depending on social, economic, and individual circumstances. Food manufacturers may very well feel unwilling for others to regulate them, and can be perfectly willing to universalize the lack of regulation. But food consumers tend to be unwilling for their food to be unregulated, and *are* willing to universalize *that* sentiment. Logic doesn't seem to be an adequate instrument to resolve this or other similar controversies. If ethical belief formation were as simple as a nail, logic might be our hammer. But if ethical belief formation is more complex than a nail, then we need more complex tools.

7.2 The Place of "Agency": Gewirth's Lockean Extension of Kant's Reasoning

It will be helpful to consider just one more example of an attempt to establish ethical principles purely by using logic. The more recent philosopher Alan Gewirth (1980) takes up this challenge in an instructively straightforward way. Gewirth thinks that what needs to be added to the Kantian analysis of moral emotions is an understanding of the notion of *"agency."*

An "agent," in Gewirth's sense, is someone who can plan on the basis of predicted consequences of actions; an agent can act rather than merely react. Gewirth's analysis of the implications of the idea of agency takes a page from John Locke (1688). Locke's treatment of what we nowadays would call agency is an often under-appreciated component of Locke's argument for the three unalienable rights of "life, liberty and property." This short list is not as arbitrary as it seems. To be *alive*, as opposed to inert and inorganic, means to be capable of *action* – relatively unencumbered by force from (currently) external constraints. Hence the need for liberty. But the freedom to act would be meaningless without the possession of some *property*, which secures our *power* to act.[4]

Gewirth assesses what is "necessary" for agency by building on the distinction between "rights" and mere "benefits." "Rights" are stronger than mere "benefits" because rights are benefits that not only would be *good* to have, but are absolutely *mandatory* benefits. It might be good for me to have a diamond ring, but I can't reasonably assert a correlative duty on the part of others to *provide* me with the ring. Rights necessarily impose duties on others, if only negative duties not to violate the rights.

Gewirth thinks the very fact that certain kinds of opportunities and resources are absolutely *necessary* (mandatory) in order for a person to function as an "agent" is enough to prove that people must assert that they have a "right" to those specific opportunities and resources. I can't avoid asserting the "right" to things that are necessary for my existence as an agent. Unlike something that is merely beneficial, a *necessary* benefit is a mandatory benefit, and thus a "right." For example, in order to be an agent in this sense, I need to be free from slavery or extreme oppression. But once I have asserted this right for myself, I must admit that it applies equally to all other agents, because I am asserting that it is a "right," not merely that I personally would *prefer* to have it.

The consensus among Gewirth's critics is that the kinds of rights established in this way would be so minimal as to be very unsatisfying as a guide to ethical conduct, because there are so few benefits that everyone would agree are absolutely necessary for a person's agency, and thus are so "mandatory" that they must qualify as "rights" (for example, Kramer and Simmonds 1996; Narveson 1980; Regis 1982). In effect, Gewirth's notion of "rights" as applied to real circumstances is as malleable as Kant's categorical imperative, and for the same reason – trying to make sheer deductive logic bear a weight that it simply can't bear alone.

If asserting a "right" is defined to include the idea that all others have a *duty* to provide whatever is necessary for my "rights," then it doesn't logically follow, from the sheer fact that I need something (even if I *really, really* need it), that I must assert that others are morally obligated to provide what I need – unless the *moral assumption has already been made* that others are obligated to provide what I "really, really need." Perhaps I am obligated to provide what a beggar really needs, but the beggar's mere assertion doesn't make it so.

Here we have a classic example of why Ayer (1946, 1965) says it is impossible to prove ethical statements with logic alone. Gewirth's argument plays out in the following logical structure:

1. A "right" (by Gewirth's definition) means whatever must be asserted as necessary for a person's agency.
2. Eating (for example) is necessary for a person's agency.
3. Therefore, everyone must assert that they have a "right" to get something to eat.

This argument is deductively valid, by the definition of "right" that is being used. But the definition of "right" that is being used is so narrow that it forgets to include the stipulation that we all have a *duty* to respect other people's "rights"![5] The hidden assumption is that we have a *duty* to ensure that everyone gets whatever is necessary in order for them to function as agents, and in fact are obligated to do so. And this ungrounded assumption is in effect tantamount to the very conclusion Gewirth is supposed to be proving. The argument is a mere tautology. As J. J. C. Smart colorfully put it, such arguments attempt to "pull a normative rabbit out of an analytic hat" (Smart 1984, p. 8).

For our purposes, an equally important problem is that such arguments can be both vague and dangerous when applied to real life moral issues. In Gewirth's sense of "rights," we can't say that people even have a right to public education, since many people throughout history have functioned as agents *without* public education. If we believe that Frederick Douglass functioned as an agent prior to his escape from slavery (as he must have, in order to educate himself and plan his escape), then it would seem that even *freedom from slavery* is not necessary for a person's "agency." In general, there is no way on this basis that we could ever establish *which* rights people do or don't have. The meaning of the abstraction "necessary for agency" is completely malleable.

Moreover, there would be no way to establish, using Gewirth's reasoning, that we ought to help people in ways that they don't strictly have a "right" to demand (in Gewirth's limited definition). This type of argument can never establish that we ought to make charitable contributions, or extend unemployment compensation in times of rapid industrial change, or help the needy widows of murdered policemen. If Gewirth's argument were valid, we would be left with a sort of Frankenstein monster of a moral system – and the most *minimalist skeleton* of one at that.

7.3 Lessons from Logic-Based Systems

The structure of Gewirth's argument is strikingly similar to Kant's. When pressed by his critics, Gewirth clarifies that the crux of his argument is that we can't avoid *asserting* that we have rights, which entails (in his view) that other agents are equally in need of asserting them – just as for Kant it was the fact that we can't be *willing* for others to treat us unfairly that entails that we must treat them fairly. And Gewirth's logical inferences end up exhibiting technical problems very analogous to the problems with Kant's argument. Because Gewirth's argument rests on a tautology, it ends up being unclear what actual rules of conduct it would require or prohibit. Are we saying that Frederick Douglass during his enslavement was not an "agent?"

Both Kant and Gewirth begin with the premise that we have a need for some moral principles, and then must choose a set of principles that are logically coherent. This process is very similar to one in which, rather than trying to demonstrate the truth of moral claims, we begin by assuming that we need a "social contract" of some sort, and thus some set of rules to govern the social contract. The arguments would then try to use deductive logic to show that there is only one particular set of rules that logically *must* be chosen as the preferred set – on pain of self-contradiction. But even if we take this turn, it will still turn out that there are a variety of ways to set up a social contract that are *not* literally self-contradictory.

In short, moral principles can't be mere logical tautologies.

A century or two of trends toward cultural and religious pluralism have now made the logical leap in such arguments seem less intuitively compelling than before. We can now easily imagine wanting others to treat us in certain ways, and thus wanting the social or legal system to require people to behave in certain ways, *without* elevating such legal and social demands into moral principles. They may consist only of a "social contract" struck for the sake of everyone's self-interested expedience.

This increasingly prevalent social contract idea, however, doesn't relieve us of the need to form coherent ethical belief systems. In fact, *social, political, and even economic policies inevitably and crucially depend on moral assumptions* – for reasons that the next section briefly takes up.

7.4 "Social Cooperation" – by Whom and for Whom?

Sociobiologists often speak of the "benefits of social cooperation" as the biological basis for human morality. The hope here is that, even in the absence of absolutely "provable" philosophical principles, some of the more non-controversial social decisions could be made simply through a cooperative social or political process or through some sort of mutually beneficial agreement – in other words, a "social contract," even if a very informal one.[6] The social contract presents itself as a way for people in a society to legislate behavior, not necessarily for moral reasons, but only because (as Hobbes argued) the parties to the contract are *benefitted* when the society protects them from certain kinds of behavior, or protects their right to behave in certain ways.

This brings us to perhaps the most embarrassing failure of deductive logic. Many philosophers, including the recently influential Rawls, have believed that, even though logic all by itself can't prove ethical statements, logic is still a sufficient tool for determining what the most "just" or "fair" social contract would be, provided we have already decided we want to live according to a social contract.

Rawls uses an explicitly "Kantian" approach in the sense that he stipulates in advance that we *need* a distributively just social contract, and then sets about to prove logically what a just social contract must consist of. Rawls's argument essentially uses deductive logic to prove what the most distributively just form of social arrangement must be.[7]

In fact, just imagine that you are *a logician and nothing more.* You know nothing except how to use logic. Even from that limited standpoint, you would be able to spell out very precisely how Rawls gets from his initial premises to his substantive conclusions. It would be an *illogical* risk for a rational person in the "original position" to opt for an extremely unequal social-political system, or one that failed to protect the most important needs (some of which are so important that we call them "rights"). The reason for this is simple: In the "original position" you don't know which position you are going to occupy in the system; thus it is much more important to *avoid the risk* of extreme deprivation than to have a chance of gaining expensive but frivolous luxury goods (on the chance that you might end up at the top of a very unequal social ladder). This fact can even be precisely measured by means of the economic principle of "diminishing marginal utility," i.e., someone's first $100 will be used for the things they consider most important for their well-being. This is true *by definition.* The next $100 will be spent for the next most important, and so on. So the "use value" of a given exchange value – the contribution it actually makes to your well-being – declines as you acquire more and more.[8] This principle is simply a fact about logic, and can't be a matter of controversy.

Rawls characterizes his approach as "Kantian," and therefore is moving in the direction of coherence reasoning. But there are important limitations of Rawls's mode of argument. First, he assumes that the prejudices and biases eliminated by the original position are the *only* impediments to knowledge about what distributive justice consists of. But this is a *huge assumption* with a great many ethical claims packed into it. At the outset, it assumes that distributive justice is the only important type of justice. It also tacitly assumes that the *only* impediment to objective knowledge of distributive justice is that our self-interest biases our judgment. Granted, once we have assumed all of this, then the question becomes how to overcome what is preventing people from agreeing as to what justice is. But *what about* all those needed assumptions?

To see why this is such an important problem, suppose for a moment, as Pojman and Feiser (2005), Sher (1987) and others often argue, that *retributive* justice is as important or even more important than distributive justice. The rational bargainers in the original position will completely disregard this question. They will not care about "retributive" justice in this sense, because their only goal is to use logical reasoning to determine which distribution system is more likely to prevent them from risking the loss of very important kinds of goods – goods that logically would not be worth the risk of giving up in order to "reward the meritorious" (whatever someone takes that to mean).

Like Kant, Rawls freely admits that he doesn't achieve the more ambitious goal of determining what our fundamental ethical principles should be by means of pure logic alone. Rawls *does*, however, assign a decisive role for deductive logic. He thinks deductive logic (including the needed risk assessments) is an adequate instrument for determining what distributive justice consists of. But as James Sterba asks in his decidedly politically conservative response, *The Demands of Justice* (1980), how do we know how *ambitious* the rational bargainers are? How badly do they want the extra goods that more inequality could make possible for those who win the game, and how hard are they willing to work for those rewards? How satisfied will the *less* ambitious bargainers be with a very unequal system that rewards them only minimally since they are not willing to work as hard to strive for them, presumably because they don't *want* them that badly? To be sure, there are Rawlsian defenses against this kind of attack (see Rawls 1993; Ellis 1998). The point for our purposes is that many assumptions of the argument, both ethical and psychological, involve controversies not resolvable by means of *a priori* logical considerations.

For all the reasons we have seen play out in this chapter, the idea of using logic, or even *primarily* logic, to arrive at ethical conclusions from non-controversial first premises seems less promising than it ever has before. But what if we combine logic with something that *can* be demonstrated in a contentful way, something that is not "garbage in, garbage out" – for example, empirical facts, especially empirical facts about human nature and human emotions? Empirical statements, after all, are with some qualifications somewhat demonstrably true or false, yet they do have content.

Utilitarianism, which still remains the dominant school of political policy analysis (see Ellis 1998), is a tempting example. Let's consider the implications of that way of thinking next.

NOTES

1 Some might object that one of Kant's formulations of the categorical imperative is that we should never treat people "only as means" and not at all as ends. But this would mean I can consistently endorse slavery, or child labor sweat shops, or any number of other bad acts, as long as in the process I *also* throw a few crumbs to the victims, so that I *also* "treat them as ends" (and not "*only* as means"). The same applies to treatment of exploited workers in modern laissez-faire capitalism.

2 The central problem here is that Kant thinks logic alone can provide sufficient grounds for eliminating all competing approaches other than the categorical imperative. In effect, he is using a coherence approach (by making ethics a matter for *practical* reason), but with only one tool in the tool box – deductive logic.

3 If there is no behavior the willing of which would be logically impossible, then no specific norms are rendered self-contradictory by the categorical imperative. Thus the categorical imperative argument doesn't make any specific rule of conduct logically necessary. I discussed elsewhere (Ellis 1998) the correlative uselessness of the categorical imperative for real-life decision making, especially for policy analysis.

4 This reasoning is similar to Theodore Cross's argument in *The Black Power Imperative* (1987). The right to eat in a restaurant is meaningless without the economic resources to *pay* for the meal. Such rights are almost empty of content. The original meaning of "Black power" referred to *economic* power – the power needed to exercise one's "rights" in a meaningful way.

5 To say "Bill has asserted a right to get something to eat" is now a mere statement of *fact:* "Eating must be asserted as necessary in order for Bill to be an agent." This definition doesn't imply the equivalence of the two statements, (1) "Eating is necessary for Bill to be an agent" (a statement of fact) and (2) "I have a duty to help Bill get what is necessary for him to be an agent, i.e., something to eat" (a moral statement).

6 The more controversial moral issues, of course, would still be left unresolved. There would also be a pronounced tendency toward a "tyranny of the majority," and many other injustices and harms would result in the absence of some philosophical justification. Yet one might hope that such a system of cooperation would be better than nothing.

7 Rawls goes to lengths in a series of journal articles, and finally in his *Political Liberalism*, to emphasize that his epistemological approach is "Kantian." He emphatically reminds us, as already indicated in *A Theory of Justice*, that he – like Kant – never intended for his argument to deductively prove the truth of his notion of justice. Like Kant, he assumes for practical purposes that we already

believe that distributive justice is a morally desirable thing. But he does use deductive logic to prove what it consists of.

8 By "use value," economists mean the actual contribution to someone's concrete well-being – as sharply opposed to its "exchange value," or what the good in question could be bought and sold for. Owning a second grand piano has little use value, even though its exchange value is the same as the first piano.

CHAPTER 8

The Limits of Hedonism
Paradoxes of "Expanded Egoism"

Ironically, happiness-maximizing views like John Stuart Mill's (1863/2015) are not guilty of a naturalistic fallacy of the particular kind that we have been worrying about here. Utilitarianism doesn't tell us to maximize the general happiness only when doing so makes *us as individuals* happy – which would be a vacuous thing to say. Instead, Mill argues that we should maximize *everyone's* happiness, as much as possible.

But there is an analogous problem. Precisely because utilitarianism *doesn't* merely tell us to do what we want to do, its main problem is how to explain our motivation to do that which, given the hedonistic assumption, we *can't* want to do. As in Kant's case, the result is a similarly ambiguous way of moral thinking, whose real-life consequences are correspondingly ambiguous. Everything from extreme laissez faire libertarianism to Pareto-efficient, centrally planned economies have cited Mill as their justification.

The virtue of utilitarianism is not the sophistication or depth of its moral psychology. We have seen repeatedly that the instrumental hedonism it assumes is overly simplistic – but that isn't the main issue for our present purposes. The interesting thing to take away from philosophers like Mill is to see a path toward a coherence rather than foundationalist way of thinking about moral issues – which I want to argue is the way we actually try to think about basic moral principles in our everyday reasoning. Mill does advance us in this direction.

Hedonistic psychological approaches often are used, both in philosophy and in real life, to establish ethical viewpoints – for example, in political policy analysis. There is no controversy in observing that people and animals value "happiness" in some sense. The problem is whether this type of motivation alone can ground moral psychology or moral thinking. The emphasis on a hedonistic empirical understanding of what we value – generically labeled as "happiness" – creates a tendency to define value in the most completely "subjective" sense discussed earlier: "X has value" just means either "I value X," or more generally "Some subject values X." Moreover, "I value X" is taken to mean "X will maximize my happiness."

8.1 Why Empirical Evidence Can't Ground Ethical Thinking: The Example of Utilitarianism

The main problem with utilitarian arguments is not simply a mistaken view that hedonism is the only motivating value of humans and other higher animals. In fact, some utilitarians actually advocate a more sophisticated value theory that is derived from but not equivalent with happiness per se; G. E. Moore (1900) represents a classic example. The larger problem with utilitarianism arises in equating "X has value" with "I value X," even when the "I" that is meant here is taken to mean every "I" in the world. Such an approach fails to answer the "Why be moral?" question – one of the crucial questions of moral psychology. We saw earlier that failure to address this question is the essence of Carol Gilligan's (1982) critique of Lawrence Kohlberg's (1973, 1981, 1984) theory of moral development in children. Kohlberg's measurement instrument wrongly assesses moral development as being higher in boys than girls, because it leaves out of account the question of what is to *motivate* actual moral behavior. The same can be said for Mill's type of utilitarianism.

Mill's central argument in *Utilitarianism* (1863/2015) exemplifies the classic strategy for using empirical evidence about human nature to ground moral claims.[1] The empirical fact that people *value* their happiness entails that it really is prima facie *morally good* to promote people's happiness. The "prima facie" qualification means: "unless someone can demonstrate that some other value should take priority." But of course (utilitarians are willing to bet), no one ever *can* demonstrate such a thing. They can show only that the value they are proposing (some type of justice, for example) is instrumental toward increasing happiness or reducing misery.

By framing the argument in this way, Mill moves us a step closer to using a coherence rather than a foundationalist epistemology, although, like Moses, he is not prepared to step into the new territory himself – referring to this argument instead as a "proof" of the principle of utility (Mill 1863/2015). What he is really trying to do is not to "prove" something, but rather to show that all competing theories *other than* utilitarianism are absolutely incoherent (and therefore no competing prima facie value can ever take priority over the happiness of sentient beings).

In effect, Mill invites anyone who would like to propose a competing intrinsic value to do so. But he also tries to show that no competing value can be demonstrated to be an *intrinsic* value, because all such demonstrations will have to proceed by showing that the proposed value produces some result that *benefits people* in some way.

Like Kant and the twentieth-century behaviorists and consummatory drive theorists, Mill assumes that benefitting people means maximizing their happiness. Given this one assumption, it would seem that if someone proposes "rights," "duties," or other moral principles by arguing that they ultimately lead to human happiness, they play directly into Mill's hands. How can anyone *demonstrate* the truth of a moral principle without appealing to the claim that such principles are expected to be beneficial to people? By attempting to make such demonstrations, they are already agreeing with Mill's original claim – that benefitting people as much as possible is the only intrinsic value. Hence the now famous thought experiments (discussed by Singer 1981 and many others) in which people are asked whether they would divert a train to kill one child if by doing so they would save five or ten children's lives – and if not, then why not? If no reason can be given, then it seems that killing the one to save the ten is the right thing to do. It is the only coherent viewpoint left standing. Utilitarianism becomes the default position.

In real life, for example, the most natural way to justify punishing criminals is that by doing so we benefit society by protecting people from crime. We have already seen that neither simple "moral intuitions," nor some religious text, nor a mere cultural tradition can justify sentencing criteria for criminal justice. But to justify punishment in terms of *protecting society* makes the punishment into an instrumental value, because the one *intrinsic* value assumed here – the end toward which punishment is only a means – is still to *maximize human happiness or well-being.* In short, utilitarianism is the only moral principle that can be known, since it is the only one that is grounded in an empirical fact – that people do value happiness.

Against this argument, G. E. Moore (1900) famously pointed out that, even if it is an empirical fact that X would facilitate the general happiness or achieve some other particular result, it doesn't follow logically that X is morally right or ought to be done. Hence Moore's resulting "open question argument": Suppose X does lead to human happiness (or to *any* preferred outcome, for that matter). We can still raise the open question: "Yes, I realize that X leads to that outcome, but is X right?" If the two concepts were really *equivalent*, then to say "Yes, I realize that X leads to that outcome, but is it right?" would be like asking "Yes, I realize that X is a square, but does it have four equal sides?" To question whether happiness maximization equates to the right action is not a meaningless question. It follows that no amount of empirical evidence about what people actually desire can prove, all by itself, that something is or is not morally desirable.

Consider any Guantanamo inmate who is tortured with water-boarding and the many other torture techniques to get information that (presumably) could protect thousands of innocent people from a potential terrorist attack (thus maximizing the general happiness). If this practice could lead to vast amounts of human welfare, does that inmate still have a moral right not to be tortured? Even if we happen to believe he does *not* have such a right, it doesn't seem logically self-contradictory to assert the contrary opinion. So Moore's point implies that there is no way to *prove* that maximizing the general happiness is more important than protecting the one inmate's individual right not to be tortured.

Utilitarians' response to this problem is instructive for our purposes: "Morally right," is not being equated *by definition* with "generally conducive to happiness" (as in "All rectangles have four sides"). On the contrary, the reason for equating "right" with whatever maximizes happiness is that *no one can offer any logical argument to show* that some other intrinsic value should take priority over it. In effect, Mill answers the "open question" with an "open challenge": Prove it to me!

8.2 How Mill Hoists Us All by Our Own Hedonistic/Foundationalist Petards

Many civil rights advocates during the twentieth century *superficially* seemed to contradict the utilitarian position because, like John Rawls, they relied on arguments about distributive justice that supposedly were independent of the value of maximizing happiness. It was argued that the happiness of the greatest possible number was not necessarily more important than the rights of a few.

But Mill's point is that such arguments always depend on showing that certain groups, even though their numbers are small, are *harmed* more by an unjust policy than others are *benefitted* by it. And Mill's argument in *On Liberty* (1859/1947) is that still more people will be seriously harmed in the long run if individual rights are neglected. This means that our opinions about what is distributively just always depend on showing that they can be derived from the principle of maximizing human happiness.

Even Rawls's idea that distributive justice is determined by "diminishing marginal utility" is at bottom a utilitarian idea. It depends on the assumption that distributing goods to favor those who are in greater need will *contribute more to their happiness* than the same goods could contribute to the happiness of those with lesser need. So utilitarians argue that even distributive justice thinkers are actually utilitarians, although they don't realize it.[2]

Notice how utilitarians like Mill use the *implicit* hedonism of other moral theorists against them, by getting them to assume that all moral motivation ultimately must serve as *instrumental* to hedonistic interests in one way or another. As we have seen, Kant is a classic example of a non-utilitarian who runs into trouble by making this same assumption. Neither Mill nor Kant offers resources to distinguish between the natural emotions of "happiness" (pleasure) and other natural emotional drives like the exploratory drive and the need for "inspiration." Moral thinking is integrally enmeshed with these other dimensions of motivation more than with the happiness dimension.

Historically, the movements toward social justice and individual rights during the past century have highlighted a somewhat stark contrast between the happiness-maximizing principle and other proposed intrinsic values. If most people would be happier – even, let us suppose, *a lot* happier – with a system that systematically discriminates against a minority group (for example, the poor neighborhoods where toxic wastes are dumped "for the greater good"), does this kind of system not sometimes, at least potentially, lead toward what Alexis de Tocqueville (1835/2000) called a "tyranny of the majority"?

Utilitarians often answer this query by assuming that treating people fairly and respecting their rights is the best way to lead to long-term general happiness (for example, Mill 1859/1947). But whether this assumption is true in all cases is still another kind of "open question."

The idea that we can cheerfully calculate the "greater good" hoists us all not only by our own implicitly *hedonistic* petards. It also uses our implicit *foundationalism* against us. If we can't think of a logical proof that something has value, then (we obediently assume) such a value can't be real. Combining both of these implicit presuppositions leads to a false impression that the intrinsic value of happiness-maximization itself has been foundationally proven. But has it?

8.3 The Muddled Moral Psychology of Utilitarianism

The initial persuasiveness of utilitarian arguments depends heavily on the way utilitarians place the burden of proof on their opponents to prove that there are some basic moral principles that can't be deduced from the value of maximizing the general happiness. By insisting that opponents must "prove" the truth of any competing value claims, utilitarians are falling back on a foundationalist epistemology.

But is there a foundationalist proof of the utilitarians' own view? To say that people value happiness, therefore their happiness "has" value, strictly speaking, is not a foundational proof unless we assume that "I value X" (on the part of some "I" somewhere) is the only legitimate meaning of "X has value." But then we would be assuming a universal egoist hedonism – which actually contradicts the utilitarian thesis that we should promote everyone's happiness, not just our own.[3]

With regard to *moral psychology*, this becomes a far worse problem than the philosophical one that Moore points out. Mill's empirical premise is that each of us does in fact value our own happiness. So we might think that the conclusion would be that each of us *ought* to act in such a way as to maximize *our own* happiness. But instead, Mill uses it to prescribe that we should act in such a way as to maximize *the general* happiness. Even if we grant that his premise implies that everyone's happiness "has" intrinsic value, the prescription that I, as an individual, ought to act in such a way as to maximally promote this value seems to conflict with the empirical premise of the argument – that our only intrinsic value is *our own* happiness. In fact, Kant had already made essentially this same point against utilitarianism in his *Critique of Practical Reason.*

If Mill's empirical premise had been that we value *others'* happiness, and therefore ought to facilitate their happiness, the problem would have been worse. The argument would then have followed the structure: "It is an empirical fact that you value doing X. Therefore, you ought to do X." In other words, we ought to do what we want to do – which is just what we were going to do anyway. But if Mill wants to advise us that we ought to do things that, according to his hedonistic premise, we do *not* want to do, then he seems to be contradicting the premise of his argument, which is that the value of X is derivative from the fact that we value it (and thus want to do it).

The main problems with this way of thinking stem from the same initial problems with which Kant struggled: Like Kant, utilitarianism assumes that our "empirical" (or natural) psychology is purely *hedonistic;* and it relies on a purely *foundationalist epistemology* by playing games with the burden of "proof." You can't "prove" the truth of XYZ (some non-utilitarian moral principle) *foundationally*, so therefore there is no reason to believe that XYZ is true. But the same applies to utilitarianism itself: It can't be foundationally "proven."[4]

Utilitarianism ends with the same motivational quandary we have seen repeatedly: Despite its protests to the contrary, it not only offers no solution to the "Why be moral?" problem. It actually makes any such solution impossible.

To be sure, as Aristotle and the recent "virtue ethicists" point out – along with more naturalistic cognitive theorists like Dennett (2003) – anyone *could have* ended up wanting to work toward maximizing the general happiness if they had made different choices *earlier in life* that would have resulted in their having certain habits of mind, and therefore certain psychological preferences, *now.* But the same problem arises at that earlier point in time. If we say that they should have chosen in such a way as to lead to such and such kind of moral attitude (a "virtue"), something *already* would have had to motivate that earlier choice. And if no such motivation was present, then again, we are telling them that they should have done something that, under the circumstances, they *could not* do.

8.4 How a Coherence Approach Is Different

The exploratory drive, as we saw earlier, might offer some of the resources needed to get us out of this dilemma. The "love of truth" might lead to respect for moral truths, but only if there is some way to establish the (at least probable) truth of moral claims; obviously, it doesn't lead to respect for moral principles irrespective of their truth value.

So the move toward basing moral psychology on the exploratory drive can be effective *only if* we take an interest in how the truth value of moral arguments is assessed. For that purpose, I have been suggesting that we need a coherence rather than foundationalist epistemology; and a coherent viewpoint must take account of the inevitably hermeneutic psychology of the motivationally driven process through which we form belief systems.

Kant, Rawls, and the utilitarians teach us that the is-ought problem has to be taken seriously as a general epistemological problem to be addressed by any coherent moral psychology. I argued earlier that the exploratory drive can steer us in the needed direction by motivating ethical inquiry, precisely *because* it is a motivation that does *not* (as other natural motives would do) predetermine the outcome of the inquiry. The exploratory drive can inspire us to look for the truth independently of whether that particular truth happens to make us happy. In this case, we are motivated, not to believe that X is true, but rather to believe *whatever turns out* to be the truth, or the most reasonable assessment of the truth. But this move does little good unless there are epistemological methods of assessing what is at least more or less likely to be true.

In short, what utilitarian thinking shows us is that it doesn't seem possible for empirical evidence alone (or even empirical evidence plus logic) to prove the truth of an ethical viewpoint – not even the statement that maximizing the general happiness should be the one essential criterion for deciding on a course of action.

More generally, and for the same reasons, it seems impossible to deduce an "ought" from any empirical "is." When philosophers try to show that moral statements can be proven from empirical ones, the results always rely heavily on selecting definitions of moral terms like "good" or "ought" that render the moral statements, in effect, equivalent to empirical claims.[5]

We can see the same problem in many different examples of attempts to derive an "ought" from an empirical "is." We saw it in the case of Gewirth, who equated "I have a right to X" (a moral claim) with "X is necessary for my agency" (an empirical claim).

As another example, Alisdair MacIntyre in his early work (see the retrospective volume of essays published in 1989) tried to show that the word "ought" is derivative from the word "owe"; thus the moral statement that I ought to repay a loan is deducible from the empirical fact that I legally "owe" the money, having in fact entered into some sort of agreement or contract. But MacIntyre later abandoned this approach, because it seems obvious that the empirical fact that I am legally required to do something doesn't logically entail that I ought to do it. Slaves and serfs "owed" fealty to their masters and overlords, and in many instances they had literally entered into signed or sworn agreements to this effect, but the system that required such debts may have been an immoral system nonetheless. In contemporary life, people owe "predatory lenders" interest on loans that perhaps ought not to have been allowed in the first place, especially if based on deceptive advertisements. Moreover, even if it could be shown that we ought to repay what we legally "owe," the question would remain whether there are also other kinds of oughts – and if so, what the appropriate balance among them might be.

8.5 The Role of Phenomenological "Evidence": A Promissory Note

There is another kind of "empirical" evidence that some might think could be used as part of the foundation for ethical beliefs, although it expands the notion of the "empirical" beyond perception of the outer world, to include "inner perception." I am referring to phenomenological reflection, which the phenomenological psychologist Ameideo Giorgi called "empirical" because it is just as much based on observations as is outer perception (for example, see Giorgi 1971).

Some philosophers and psychologists tend to write off phenomenology as merely "folk psychology," a kind of self-reporting of notoriously unreliable subjective states – as when a husband angrily reports to his wife that he is not angry. Actually, phenomenology is an elaborate discipline aimed at getting *beyond* superficial readings of what we think we feel or think, to get underneath all the distorting hermeneutical prejudices and preconceptions that cause our superficial reports of subjective processes to be so inaccurate and frequently altogether mistaken. In this way, phenomenological methods are designed to get closer to an understanding of what we really think or feel (Husserl 1913; Merleau-Ponty 1945) – even if we will never perfectly understand with ultimate precision what our subjective states mean or what they "are about," as acknowledged by Gendlin (1978, 1992, 1962/1997), Husserl (1936/1970), and Merleau-Ponty (1945, Preface).

Phenomenology can't absolutely nail down the meaning of our feelings, but it can move us closer. If we apply phenomenological methods rigorously, we can come to realize that what our feelings about homosexuality really reveal is *not* that we have an "intuition" that homosexuality is "morally wrong," but a personal feeling of disgust or a fear that acceptance of homosexuality may cause our children to be more likely to become gay; we then misinterpret this fear as a moral judgment. But careful reflection can eventually reveal the nature of such feelings, if unwarranted presuppositions are separated out from the raw feeling quality. Most people can eventually learn to recognize the difference between fear and a moral judgment, if they examine the feelings in their felt quality *before* jumping to conclusions about what they "mean."

From this standpoint, phenomenology can make an important contribution toward helping us become more emotionally self-aware as a prerequisite for interpreting what our emotional feelings are telling us about ourselves and our connections to our environment, within the larger context of our ongoing hermeneutical process.

But can such introspections *foundationally ground* ethical beliefs? Can they overcome the "is-ought" problem in a way that in principle is any different from the empirical assumptions of Mill and other utilitarians, or from the intuitionism of Moore? Even with "phenomenologically refined" intuitions or emotions, we still can't simply equate an emotion by definition with "the good" (happiness or any other emotion).

If we begin with "common intuitions" and then refine them phenomenologically, we still end up with a situation in which the "refined" intuitions of different people contradict each other. The reason is that the phenomenological reflections we undergo are already guided by a hermeneutical process that motivates us to "refine" our intuitions as little as possible in order to preserve as much as possible of the initial "intuition." Thus even trained phenomenologists have very different intuitions when it comes to questions of distributive justice in a sociopolitical system. Heidegger, famously, supported the rise of Nazism – obviously contrary to the view of other phenomenologists like Simone De Beauvoir or Hannah Arendt. And such a classic phenomenological ethicist as Max Scheler insisted that the main purpose of punishment is retributive desert, not deterrence or rehabilitation, whereas many others disagree.

This impediment remains in effect as long as we try to use phenomenological methods as a *foundationalist* epistemological justification for beliefs. But in the context of *coherence* epistemology, the contribution of phenomenology will become crucial. In the final analysis, if coherence criteria are our best hope for an epistemological basis in ethics, then the clarification of the meaning of emotions is going to be an essential part of the process, for two reasons: First, because emotional motivations are the key to why we tend to distort evidence and arguments in order to prop up an overall hermeneutical worldview. These distortions subtract from the coherence of a belief system.

And secondly, as we have already seen, the emotional reading or misreading of the implications of SEEKING, CARE, and the many other emotional systems and their inter-relations partly determines what is possible for moral epistemology. It is tempting to categorize ethical impulses as manifestations of only the CARE and other "social-bonding" emotional systems. But careful phenomenology reveals that the situation is much more nuanced than this, and requires the participation of the SEEKING system. What careful phenomenology can reveal most of all is the ways in which motivations including but not limited to SEEKING interact to affect in each instant the shape of the hermeneutical process for both better and worse.

From a foundationalist standpoint, phenomenology alone can't get us to reliable and practically applicable moral principles. But for purposes of coherence evaluations, apodictic certainty is not the goal. In real life practice, we are only trying to determine whether one moral viewpoint is more coherent than alternative viewpoints. Crucially, this goal requires understanding how emotion affects the hermeneutical process.

Misreading of the meaning of our own emotions may well be a leading cause of incoherence in a person's moral outlook. Many people, for example, can't *imagine* any understanding of emotions in which the ultimate goal of all emotions is not to maximize rewards. Their reading of what they *mean* by value terms implies that all emotions are hedonistic by definition. Yet, as we have seen, neurophysiology now suggests that there are non-hedonic motivations – an idea that is consistent with reservations that phenomenologists have been expressing toward hedonistic psychology for some time now. Until now, psychology has mostly regarded non-hedonistic behavior as an aberration and a paradox to be solved – just another occasion for Thomas Kuhn's "normal puzzle-solving."

There is still another crucial need for careful phenomenology. We can't know the "what it's like" correlates of a physiological process by empirical methods alone; in order to say "this physical process equals pain" (in the experiential sense of "pain"), we already would have to know what "pain" feels like. Similarly, as Panksepp warns, emotion systems like CARE and PANIC should not be equated with any preconceived category of *conscious* emotional feelings. The conscious dimension can be studied only with the help of phenomenological introspection. This is only one simple example of why phenomenological techniques for clarification of the meanings of emotions becomes a crucial part not only of moral psychology, but also moral epistemology.

Let's take stock of where the discussions up to this point have led us about the current dilemma of moral epistemology, especially when we bring contemporary psychology into the picture.

8.6 The Is/Ought Dilemma and the Inevitability of Coherence Reasoning

When philosophers seem to be arguing about what the "true meaning" of "good" and "right" are, they hopefully are not engaging in a silly dispute about which definition they choose to pull out of the dictionary, and insisting that other definitions should be ignored. Instead, they are arguing about whether a definition can be used *in the context of a viewpoint that has a chance of being supported with good arguments.* When Moore objects to Mill's defining "good" to mean "valued by people," he is not merely insisting on a preferred definition, as someone might insist that "democracy" should be used in one way rather than another. What he is saying is that Mill should not equivocate the term, nor ignore the fact that there could be other senses of the term that could also be used in meaningful ethical claims that don't fit Mill's theory.

On the other hand, neither is Mill merely insisting on a preferred definition of "good." What he is claiming is that he can use his definition in a theory *that can be grounded in reasonable arguments*, whereas other competing definitions, while they can be used in ethical claims, *can't* be used in ethical claims *that can be grounded in any reasonable arguments.*

So neither Mill nor Moore is falling into the elementary fallacy of quibbling about what words "really" mean. The real question is not whether we can show that one definition is "better" or "closer to what 'we' really mean" than another. Part of what we should learn from the disputes between Kant, Mill, and Moore is that arguments that depend on defining a term in a certain preferred way don't get us very far. They lead to belief systems that are tautologous and therefore trivial.

But if this is the case, and if those disputes remain as unresolved as the above discussions would suggest, then where does that leave us? Let's now see if we can establish that there is a viable "coherence" method, as opposed to foundationalist "proofs" of ethical beliefs. In real life, people can't rely on foundational proofs any more than they can on "moral intuitions," or supposedly un-interpreted religious texts, or supposedly relativized cultural traditions. How, then, can people form moral viewpoints that they believe to be actually more reflective of the truth than alternative viewpoints? This question will be explored next.

NOTES

1 Singer (1981) adds a new argument: society as a whole will not *accept* any proposal that doesn't count everyone's interests the same. But there are many candidates for what "the society" will accept as a reasonable argument (the Rawls and Sterba viewpoints, for example). This diversity of what is "socially acceptable" forces utilitarianism again to fall back onto Mill's initial basic argument.

2 Mill can argue that distributive justice derives from the *utilitarian* principle that we should distribute resources where they do the most good – i.e., according to the principle of diminishing marginal utility. Rawls's formula merely maximizes well-being while correcting for diminishing marginal utility (e.g., Ellis 1992, 1998): The same goods have more *utilitarian* value in the hands of those who need them most.

3 The utilitarian assumption is that justice will lead to long-term general happiness; but this is a questionable assumption – and it also seems to implicitly acknowledge that justice is a *prima facie* value. After all, why fall back on this *ad hoc* empirical hypothesis unless it is already assumed that justice is important

in its own right? Why not just say "Whenever justice doesn't serve the general happiness, it has no value"?

4 This divide between "is" and "ought" becomes still more stark in light of contemporary psychology – which highlights the idea that we *can't* do any behavior that we aren't *psychologically motivated* to do. We saw earlier that this problem is not avoided even if we assume that human nature is not necessarily inherently selfish.

5 Some have even argued that it is merely a linguistic trick for Mill to infer from "what people actually value" to "morally valuable," and from "what people actually desire" to "desirable" (Ellis 1998, 1992).

CHAPTER 9

The Hermeneutic Process in Action
Fallibilism and the Role of Emotion in Moral and Political Psychology

9.1 The Dangers of "Anything Goes" Relativism

We are no longer naïve to the fact that different people interpret the same text or tradition in utterly different ways. This modern and "postmodern" trend interacts with an increasing acknowledgement of the difficulty of philosophically grounding ethical belief systems, as the previous two chapters illustrated. The result has been a tendency during the past century, in both academic philosophy and the general culture, toward various forms of ethical relativism.

But ethical relativism is not at all a satisfying resolution of the dilemma, and can be dangerous in real-life terms. If people believe that ethical statements are neither true nor false, but only expressions of emotion, or of social conventions and cultural customs, then one of the big dangers is that at this point "anything goes." Fascism or genocide can just as well be grounded in an emotional preference or social convention as can democracy, generalized benevolence, or human rights.

Worse still, if those who lack economic or military power are content to allow social or political policies to go without the need for arguments to justify them, then those who *do* have economic or military power will always win the day. The less powerful lose even the ability to persuade others to join their cause on the grounds that it is morally right – as abolitionists persuaded many citizens to fight to abolish slavery in the United States. If no moral arguments can be made, then this force for social justice or even social beneficence is lost. Every policy recommendation would become just as "valid" as any other. Every destructive behavior would become as morally defensible as any other.

So let's consider whether perhaps relativism and foundationalist proofs are not the only alternatives. Perhaps we are suffering from a self-imposed *false dilemma* when we think that we must either be able to prove something with a great degree of certainty, or not at all. Is there a way to establish some degree of epistemic grounding for ethical belief systems, although without absolute certainty?

It can now also be argued that a kind of "coherence" epistemology fits nicely with the new understanding of basic emotions in which the exploratory drive is relatively independent from other needs and wants. So we can see that there is at least some genuine incentive *not* to deceive ourselves with hermeneutical casuistry.

Up to this point, we have viewed hermeneutical distortions of belief systems almost as though forming true beliefs could have been easy if not for those distortions. But that is not the case. Moral thinking at best is too challenging for such a simple opposition. We have already seen why it is inescapable that we must accept a healthy degree of epistemological skepticism. Not only does this imply the need for a coherence rather than foundationalist approach to moral epistemology. We are also beginning to see, from the above historical excursions, why the coherence approach itself is more difficult than one might initially suppose.

Moreover, the uncertainty of moral reasoning *interacts with* the hermeneutical distortions to which we are always tempted. The greater the degree of uncertainty, the greater is the temptation to use that very uncertainty as an excuse to believe whatever we prefer to believe.

One of the main purposes for studying the history of academic philosophy is to learn from its mistakes. We reviewed somewhat extensively several examples of philosophers who try to "prove" that their particular brand of ethical system is the only coherent approach. Everyday people often fall into this trap as well. We all long for the certainties offered by a comforting geometric proof. The lack of such "proof" triggers anxiety, moral ambivalence, appeals to authority, and the temptations of self-deception.

But a more constructive lesson from the limitations of foundationalist "proofs" is that we need a different strategy for moral reasoning. What our historical examples showed is that an effective coherence strategy demands the inclusion of more than just deductive logic combined with empirical evidence. We need more extensive criteria by which to evaluate the coherence of competing valuational viewpoints.

When utilitarians add empirical evidence about human psychology to the epistemological toolkit, their approach can be seen as a coherentist strategy expanded slightly further. Part of the problem here is insufficient understanding of emotional processes, but another problem is the foundationalist approach to epistemology per se. Looking back at all these philosophical theories, it now seems obvious why moral thinking requires a "big-tent coherence" epistemology. By "big tent," we mean that the coherence criteria must include still more even than deductive logic combined with empirical evidence.

In our everyday lives, coherence epistemology is actually the way we normally think. We begin with a hermeneutical belief system that has previously been adopted, for either good or bad reasons, and then look at anomalous facts that tend to challenge our overall view of how reality works. At this point, we don't necessarily just engage in the casuistry of trying to eliminate self-contradictions within the original viewpoint. Instead, we *look for a better viewpoint altogether.* Just as in Hegel's historical progressions in which a "thesis" is opposed to an "antithesis," the new "synthesis" can't just be a tweaking of the thesis or the antithesis, nor is it a mere compromise between them. It must be a new viewpoint capable of avoiding the coherence problems of both the thesis and the antithesis.

We use coherence criteria in trying to assemble a new belief system that makes as much sense as it reasonably can, all things considered. Or if we try to *refuse* to engage in such a thought process (for example, if we are authoritarians, extreme conformists, or extremely incapable of introspection), then we merely "explain away" the anomalous facts, and end up with an increasingly incoherent valuational outlook, which leads to further inner turmoil that must be addressed with even more incoherent thought processes. This situation inexorably leads to a *less* coherent value system. So our next step should be to explore the way a coherence strategy can work when it is working as effectively as it can.

9.2 The Need for a Big-Tent Coherence Epistemology

In the context of all the epistemological challenges we have discussed here, there are several major hurdles that will always erect themselves against every attempt to ground a non-trivial system of ethical beliefs. One, as already mentioned, is the is-ought problem, whose ultimate challenge is the danger of committing the naturalistic fallacy – asserting that I ought to do X because I want to do X, and thus in effect I ought to do whatever I want.

But on the other hand, if we *don't* ground ethics in psychology, we end up asserting that people ought to do things that they are incapable of being psychologically motivated to do. We began to lay the psychological groundwork for escaping from this trap in the earlier chapters, but we still must worry about how to separate out the results of truth-seeking from the results of *hermeneutically motivated distortions* of reality.

Another major problem, which actually applies to every proposed epistemological grounding, is the problem of circularity. Whatever epistemological grounding we propose, the question can always be asked why we believe epistemological principle X is the best one to choose. If the epistemological system that grounds our philosophical beliefs is X, then we are claiming that we believe X because we believe X – which obviously is circular.

We saw this problem clearly in the case of Ayer's emotivism. By assuming that only logic and empirical evidence can be used to ground beliefs, he concluded that moral statements can have no epistemic validity at all (since they can't be proven in those ways), and therefore are only expressions of emotion. But a major problem with Ayer's argument was in his assumption that logical empiricism is the only acceptable epistemology to begin with – especially since the "empiricist criterion for meaning" itself is not empirically verifiable. In general, if I ask "why should I use X as my epistemological grounding?" the answer must already presuppose an epistemological grounding, and so on ad infinitum.

There can be no final answer to the problem of epistemological circularity, but the low-hanging fruit in this problem is to consider some of the obvious causes of incoherence in the hermeneutical process of forming and maintaining belief systems. Considering the interplay between the epistemological difficulties we have discussed and the hermeneutical process, we can identify at least five major considerations that will affect the coherence of a person's moral outlook, with the fifth one outweighing the others in importance:

i. *Logical consistency.* It would be impossible for any epistemological approach not to at least include a principle of non-contradiction, in the strict or narrow sense of that term. In point of fact, I have never read any philosophical or social-scientific passage, even the most epistemologically skeptical (from Hume to Foucault to even Zen Buddhism), that did not make deductive inferences. It is difficult to imagine how any proposed epistemology could serve as a criterion for distinguishing between good and bad beliefs if it allowed that the good ones might just as well entail their own (literal) negation or vice versa.

On this assumption, consider the following thought experiment. Suppose Kant's hoped-for outcome had been true: suppose, of all the proposed theories about something, *all except for one* were to lead to logical self-contradiction. Then that remaining one could be taken as the best viewpoint, provided that the array of possible viewpoints being considered is exhaustive.

Now of course this array of possible viewpoints can seldom if ever be assumed to be exhaustive, and even if it were, it is very unlikely that all but one theory would lead to a contradiction – as we saw when considering the examples of Kant, Gewirth, and Rawls. So logic alone can't distinguish coherent from incoherent moral beliefs.

ii. ***Consistency with established empirical facts.*** The fact that no single viewpoint is completely provable through logical non-contradiction is actually what makes a coherence approach most necessary. Suppose we are choosing from moral opinions A through Z. It often will *not* turn out that all the viewpoints from A through Y lead to logical self-contradictions (which would suggest that Z is the only viable viewpoint). Instead, it may be that some of the belief systems, although internally logically consistent, nonetheless contradict verifiable *empirical facts* (assuming that we have developed probabilistic criteria for assessing these facts). Then those moral systems too, because they contradict likely empirical facts, can be regarded as unlikely to be true. However, as we saw in the discussion of utilitarianism, even logic plus empirical facts still can't provide an adequate basis for choosing between competing moral viewpoints.

iii. ***Parsimony.*** The term "parsimony" can be misleading, and should not be used to justify overly simplistic belief systems. What the "principle of parsimony" really means is that ad hoc hypotheses should be avoided as much as possible.

Why does this principle add to coherence? Suppose that some proposed viewpoints, while not contradicting either themselves *or* well-established empirical information, nonetheless require accepting without any evidence many more *ungrounded assumptions* than theory A does. This would be a reason to think, purely on statistical grounds, that A is more likely to be true than those other theories (all else being equal). This is the case for two reasons. In the first place, the more ungrounded assumptions there are, the greater the probability that one or more of them is false.

But more importantly, if we allow someone an *unlimited number* of ungrounded assumptions, they can defend *any view whatever*, regardless of how absurd it is. As a simple example, consider any elaborate "conspiracy theory" – for instance, the theory that President Barack Obama was not really born in the United States. At the time of this writing, a surprisingly large minority of the US population still believe that this theory is true, despite its eventual disavowal even by its loudest proponent, Donald Trump. Some of the facts with which this theory must be reconciled are the following: The Governor of Hawaii, who was a Republican (and thus not particularly an Obama supporter), examined the official birth document and confirmed that it was valid; Obama later released to the media an official birth certificate, which was openly displayed and reproduced in the press; furthermore, a Hawaiian newspaper had announced the birth of Barack Obama in Hawaii on the date of his birth.

Now suppose we allow the proponent of the conspiracy theory an unlimited number of ungrounded assumptions (or, in scientific terms, "ad hoc hypotheses"). All of the conflicting data just listed can be explained away. We could assume that Obama's birth certificate was a forgery; that the Governor of Hawaii was lying. As for the newspaper announcement, it could be explained by the presumed fact that Obama's mother (whose US citizenship automatically made her son a citizen, regardless of birthplace) sent a telegram from Kenya asking that the announcement be published, because she was already planning that, forty years later, her son might want to run for President and therefore would need to be a "natural-born" US citizen – even though there is no evidence that such a telegram existed. Finally, we could assume that a "liberal media bias" predisposed the members of the news media who were allowed to view the official birth certificate to believe that it was authentic, even though in reality it was a forgery. And of course there would need to be an explanation for why a woman who verifiably lived in Hawaii would choose to travel to Kenya to give birth, forcing herself then to go to great lengths to make it appear that the child was born in Hawaii after all. Presumably, some ad hoc hypothesis could explain this fact.

If granted all these ad hoc hypotheses, for which there is no evidence, but which nonetheless are not logically impossible, the conspiracy theory can be defended. But the *more* such ad hoc hypotheses are needed in order to make the theory fit the facts, the *less likely* the theory is to be true – all else being equal.

The nature of this conspiracy theory, in a nutshell, illustrates the real basis for the "principle of parsimony" in epistemology and philosophy of science. The principle of parsimony is often justified by citing "Ockham's razor," which states that if a simple theory and a complicated theory are equally capable of explaining all the facts, and are equally credible in all other respects, we should choose the simpler one.

But let's be more clear: The fact that a famous medieval philosopher said something is not a reason to believe it. The real reason is the fact that it is easy to justify *any* theory, no matter how preposterous, if allowed *enough* unfounded assumptions. The more unfounded assumptions there are, the easier it is to make a false theory fit the available facts of the case. Conversely, the fewer unfounded assumptions a theory must make in order to make it fit the facts, the more likely it is to be true, all else being equal.

Consider a specifically ethical example: Todd Aiken, a member of the US House of Representatives, believed that abortion should be illegal in all cases, but he also believed that it was morally problematic to force a woman to give birth to the child of her own rapist (Guardian 2012). As a solution to the dilemma, Aiken adopted the ad hoc hypothesis that a woman who is raped experiences so much stress that the stress prevents her from becoming impregnated. He cited one "fringe science" study (which contradicted many studies in more reputable journals) that would be vaguely consistent with this hypothesis, which not only required him to greatly exaggerate the results of the study, but also required the additional assumption that there is no substantive *methodological* difference between real science and fringe science. These additional ungrounded assumptions about the nature of science would then require still further assumptions in order to defend them against points made in standard textbooks on scientific method, and so on ad infinitum.

Without as many ungrounded assumptions, it is more difficult to maintain a false view. Conversely, the *more* ungrounded assumptions we are allowed, the *easier* it is to maintain a false view. It follows that, all else being equal, a theory that requires many ungrounded assumptions is less likely to be true than one that requires fewer ungrounded assumptions.[1]

On the other hand, as I warned earlier, if we were to give the principle of parsimony too much priority, we might prejudice our thinking in favor of overly *simplistic* systems of thought. In the context of ethical thinking, however, thought systems that initially seem simple actually turn out to be intricately complicated when they play themselves out in the full context of a person's overall view of life.

This parsimony issue is one of the main reasons why people have developed ethical *theories* and value *systems* as opposed to random lists of moral rules of conduct. If the rules of conduct can be inferred, even imperfectly, from a very short list of initial assumptions, the resulting coherence of the overall belief system causes us to see those rules as more plausible – at least in this one respect – than if they were merely in a random list.

So avoiding proliferation of ungrounded assumptions can be taken as at least one criterion for the comparative coherence of moral viewpoints, along with the competing viewpoints' consistency with reasonably established empirical facts as well as internal logical consistency.

No single incoherence completely eliminates a viewpoint from the running – not even internal inconsistency. After all, most versions of quantum theory lead to at least a few self-contradictory implications, but this by itself doesn't cause scientists to abandon the theory; it only causes them to look for ways to improve the theory.

iv. *Avoiding bad empirical reasoning: Similar basis for parsimony and for the probability of empirical facts.* The criteria for what should count as a "reasonably established empirical fact" must account for the problem that there can be competing explanations for any set of data. Suppose we claim that a reliable correlation between bible sales and liquor sales establishes that religiosity drives people to drink. Obviously, there is a strong probability that one of a thousand other competing explanations for the correlation could be the correct one; thus the one we have proposed is probably false.

By contrast, suppose we assert that a correlation between child abuse and later violent criminal activity on the part of the abused child implies that child abuse is one cause of violent criminal personality development. In this case, there are *fewer* possible competing theories, and most of them can be eliminated by introducing the needed controls into our empirical studies. We can account for a possible genetic explanation by appeal to studies of adoptive children, identical twins raised together, twins raised separately, etc. In this way, we can show that there are fewer plausible competing explanations, thus increasing the probability of the truth of our hypothesis that child abuse is a cause of later violent criminal development, at least in some cases. "Multiple regression" statistical techniques are designed to achieve these kinds of "well-controlled" results.[2] Ethical theories that require contradicting probabilistic "facts" rooted in decent empirical methods of this kind would therefore *lose some of their coherence*, all else being equal.

How does the reliability of empirical findings play into the coherence of *moral* belief systems? As a simple ethical example, suppose someone believes that homosexuality is morally wrong, yet does not want to blame people for character traits that are genetically inherited. If it turns out as an *empirical fact* that homosexuality is largely genetically inherited, this will pose a severe coherence problem for such a moral belief. Indirectly, this in turn can pose a coherence problem for the overall belief system from which the condemnation of homosexuality is derived – for example, a belief that the presumed truth of Biblical scriptures can provide a final answer to all kinds of moral questions. If we begin by doubting the Biblical claim that homosexuality is immoral, we eventually may find ourselves also questioning even the simple moral principle that it is wrong to boil a kid in its mother's milk (as clearly stated in Deuteronomy 14: 21). It could also cause us to question the Biblical principle that we can cure hemorrhoids by erecting golden statues of hemorrhoids in the temple (1 Samuel 6:7). So our motivations to deny the empirical fact that homosexuality is genetically inherited *multiply* as its various *implications* play out in context.

We must always allow for the possibility that alleged "facts" may later be disconfirmed, just as the medieval theories of chemistry and physics were. Yet we are more likely to choose a better moral viewpoint if we choose one that on the whole coheres with facts that have been reasonably well-established as opposed to choosing one that conflicts with many such facts. To be sure, cohering with such facts is not a final or decisive criterion, and all the less so when the implications of the findings have been couched in ambiguous language – as unfortunately often occurs in the "analysis and conclusions" section of empirical studies, where the researchers may want to hedge their bets as much as possible, and therefore avoid words like "cause" and "effect."

Here again, no *one* such incoherence would completely take a viewpoint out of the running; but a theory that contains many such incoherencies would become correspondingly more dubious.

v. *Coherence with phenomenological facts.* As indicated earlier, phenomenological facts can play a role in coherence criteria similar to the role of empirical facts. A simple example is the phenomenological experience of "freedom." Are we free only from currently external forces, or are we also free from the environmental and hereditary factors that may have caused us to become who we are?

These are two completely different senses of "freedom." We subjectively experience the fact that our mental choices can cause our bodies to move, and thus indirectly affect other physical objects. If we deal with this question by saying that introspectively experienced mental events have no real causal power to move our bodies – that their causal power is "illusory," as Wegner (2003) maintains (with Dennett's 2003 mildly qualified endorsement),[3] then we are faced with many logical incoherencies. For example, we can be reasonably certain that all physical events have causal power. If we then posit that mental events have *no* causal power, we are logically implying that mental events are *non-physical* events. But then we are endorsing a metaphysical dualism of the kind that became so problematic for Kant. The idea of "non-physical" realities entails its own coherence challenges.[4]

So the *phenomenological experience* of "freedom" – that our mental processes play a role in moving our bodies, regardless of whether we ourselves were predetermined to be who we are – is a phenomenological datum that must be accounted for in a coherent worldview. On the other hand, coherence requires that we not conflate this kind of "freedom" with the contra-causal sense of "freedom." Willingness to acknowledge this distinction also requires phenomenological honesty: The narcissistic issues that we all must grapple with may get in the way; we would all like to be more in control of our situation than we actually are.

Consider a specifically ethical example: Suppose I can construct an ethical viewpoint that *would* be coherent if I assume that "most cultures have disapproved of public nudity." But then suppose that, when I engage in more careful phenomenological reflection than often is done, I realize that my own presumed disapproval of public nudity is actually not a moral judgment, but only a personal feeling of disgust as in a child's disgust with eating asparagus. And furthermore, suppose I notice that the sequence of phenomenological clarifications that took place in my reflective process are not contingent on any particular idiosyncrasies of myself as an individual. In other words, anyone with a similar starting belief could have asked themselves *the same questions* that I asked myself. It now seems likely that any other person who thinks they morally disapprove of public nudity, and whose ostensible reasons for the disapproval are the same as mine had been, would therefore most likely come to understand the illusory nature of those reasons in the same way that I did. Now I *can no longer feel confident* that "most cultures have morally disapproved of public nudity." The phenomenological clarification shows that cultures in fact have not *morally* disapproved of public nudity at all. They were only confused about their own emotions and the reasons for them.

A similar phenomenological clarification often takes place in the minds of criminal justice students. Initially they feel that "moral desert" is the basis for punishment, and that most other people feel the same way. But after immersing themselves in the details of criminology, they often begin to realize that what they really felt, even in the beginning, was that punishment is necessary because of its deterrent effect – not because of the offender's "moral desert." Others, of course, become more clear that "moral desert" *is* the basis for their position, but they also become more aware of the difficulty of establishing the truth of such a position.

The carelessness with which people phenomenologically reflect on the intentional meaning of their supposedly "moral" feelings is actually one of the most important causes of incoherence in ethical viewpoints. But as we have already seen, to incorporate phenomenological methods into a larger coherence strategy is by no means to advocate that phenomenological reflections can reveal "moral intuitions" as an adequate basis for ethical beliefs. Even a phenomenologically clarified intuition is not an adequate reason for believing that the resulting intuition reveals a statement about the world as opposed to a subjective preference.

The hermeneutic circle itself makes phenomenological reflection difficult and vulnerable to the unconscious foibles of self-deception. This is why Stephan Strasser (1969) argues that phenomenology should be "dialogal." A diversity of voices can help guard against some of the unwarranted and unconscious presuppositions of any one individual, which may remain invisible to that individual until diverse voices call attention to them – provided that the diversity *really is* diverse.

In sum, the overall coherence strategy is that by using internal logical consistency, parsimony, and coherence with reasonably established facts as criteria for the coherence of moral perspectives –both empirical *and* carefully clarified *phenomenological* facts – we can compare competing viewpoints and choose the one that, all things considered, is the most coherent.

In practice, especially in our everyday moral reflection, the carefulness of our phenomenology may well prove to be the single most *important* component, given the effects on ethical incoherence of careless introspection, as we can easily see by observing earlier periods of history. When we look back to earlier periods, their distorting prejudices and preconceptions seem more obvious, not necessarily because they are less veridical than our own, but only because they are different from our own.

As Merleau-Ponty (1945) suggested, and later hermeneutic philosophers fleshed out in more detail (for example, H. G. Gadamer 1971/1975; Gerd Brand 1967; Ludwig Landgrebe 1966), no phenomenological reflection can completely eliminate all distorting preconceptions from our interpretive process. Phenomenology will always be an ongoing struggle; it needs to be approached in the same spirit of fallibilist coherentism as any empirical inquiry. But its uncertainties should no more disqualify it a priori than do the uncertainties of science – as one *component* in a larger coherence strategy.

NOTES

1 Given ten thousand ungrounded assumptions, I could deduce a dozen theories that account for all relevant facts in a given problem area. It is much less likely that I could deduce even two or three theories that are consistent with the facts if granted only *one or a few* ungrounded assumptions as a starting point. So a theory requiring fewer ungrounded assumptions is more likely to be true than one that requires ten thousand ungrounded assumptions.

2 The assessment of empirical facts is dependent on the *im*-probability that a correlation would occur just by chance or coincidence, combined with the completeness with which competing explanations are ruled out.

3 See Ellis (2013) for further discussion.

4 For example, can a non-physical thing be in a physical place in space? If not, how is it affected by *physical* forces in the brain? Why should *your* consciousness always be affected by *your* physical body rather than being affected by mine? It seems more coherent that mental processes have some of the same causal powers that their own "physical" correlates have. Then they would possess causal powers, as all other physical processes do.

Conclusions

Naturalistic approaches to moral psychology in recent decades have made slow progress, in my view, precisely *because* they have focused primarily on the naturally altruistic and cooperative sentiments. This emphasis, as we have seen, tends to trivialize ethics by trapping it into the naturalistic fallacy, which would reduce ethical thinking to a listing of psychological facts. The additional emotional dimension needed is the idea of an "exploratory drive" as a primary motive (as opposed to a learned behavior pattern). In the twentieth century, an independent and primary exploratory drive was considered as at best hypothetical, and was treated with considerable skepticism by behaviorists and most drive theorists. But we now have made more progress in understanding the neurophysiology and phenomenological experience of the exploratory drive. In humans, with our vastly developed prefrontal thinking and self-regulatory capacities, including the voluntary control of mental imagery, the drive to explore reality includes the realms of value and meaning.

The occasional (or even frequent) feeling of altruism is not enough to motivate ethical thinking and behavior. The KKK Grand Dragon mentioned earlier often acted out of altruism, but he could also be cruel when his racist belief system was at stake – as well as other, more basic aspects of his overall worldview, the questioning of which was anxiety-provoking. In addition to altruism, ethics also requires that there be an independent desire to discover the truth – independent enough that it functions even in the face of a persistent fear always lurking beneath the surface: fear of the danger that the truths we find may undermine a previously accepted worldview, and may not necessarily be in the service of other needs and desires. We implicitly fear that, if our overall view of reality turns out to be fundamentally built on quicksand, then we may be forced to conclude that everything we have done in our lives is correspondingly wrong or dysfunctional.

We have seen that, unlike other, more "social" emotional systems, the SEEKING system does not predetermine *what* ethical feelings, beliefs, or behaviors we are going to end up with. This requires some rudimentary intellectual work, which however must be *motivated* by the SEEKING system, independently of the various intrinsically truth-indifferent emotions. The SEEKING system merely determines *that* we will try to find out the truth as best we can. If all motivations were to collapse the "that" into a "what," this is where the naturalistic fallacy would block the possibility of non-trivial ethical thinking, because in that case our emotions would determine *what* moral principles we adopt, leading to a trivial moral system: "Do whatever you want."

For this reason, many seemingly easy ways out of the "Why be moral?" dilemma don't work out well as long as we are stuck in a motivational approach that reduces all moral commitments to naturally altruistic sentiments or, still worse, to mutually egoistic cooperative tendencies – regardless of whether those are innate or socially learned. If altruistic behaviors are motivated only in those ways, their effect on ethical validity is severely limited by the naturalistic fallacy problem.

Some might insist that social approbation provides a sufficient motivation – for example, that we want to be moral, in effect, in order to look good to others or win their approval. But we have seen that this view would place an unrealistic burden on the "nurture" side of the nature/nurture interaction in moral psychology. The natural motivational systems that provide the basic equipment driving moral development need to include (and do include) a good measure of nature, not just nurture. The development of the "neo-amygdala," as we saw earlier, is a spontaneous *pattern of brain growth;* the "old amygdala" doesn't dictate a "brute, egoistic nature" that has to be overridden by social conditioning. Social conditions, of course, can affect the natural patterns of moral development by nurturing or failing to nurture them. No matter how much innate tendency a plant has to grow, it still must at least have light, water, and a few essential minerals. Other than rare and extreme sociopaths, we all have the equipment that drives both altruistic concern *and* the search for general truth on such matters – with two important qualifications: that this motive can conflict with our own competing motivations and interests, and that it requires certain opportunities and resources in order to express itself.

It would be too facile to try to escape the naturalistic fallacy problem by saying that moral actions are a *subset* of what we naturally want to do, in conflict with other motives that are "equally natural." In the absence of an unconditioned exploratory drive leading to truth-functional ethical beliefs, we would still end up doing all and only those behaviors that we are *most strongly* motivated to do. We would still behave morally when we want to, and immorally when we want to. The crucial problem would remain, that moral development requires a distinction between simply a motivation to be nice whenever we feel like it, and on the contrary a motivation to find out when and in what ways we *ought* to be nice.

Even a hedonically learned *non*-hedonic behavioral habit (for example, following the norms of one's culture because of social conditioning) provides the basis only for descriptive, not prescriptive accounts of behavior. "You should act in the ways you have been socially conditioned to act" provides no basis for evaluating how we *should* be socially conditioned to act – nor, for that matter, *which* of the various conflicting social influences we should pay attention to. If we say that we should socially condition our children to be religiously tolerant, we are assuming that we, the previous generation, could have had some *motivation* to condition them in that way in the first place – that we want to do so. In other words, we are still saying that everyone should do what they (or the previous generation) already wanted to do.

This is where the exploratory drive and its selective suppression become the pivotal concerns. Ultimately, the main focus becomes how the conflict between the love of truth (motivated by the exploratory drive) and other natural emotions gets resolved. This conflict affects my selective inattention and thus the way I resolve conflicts within my philosophical worldview. The coherence of my worldview depends on how these internal conflicts are dealt with.

The fact that we intrinsically value the well-being of certain specific others in specific circumstances wouldn't get us out of Moore's naturalistic fallacy, until we combine it with the naturally human intellectual work motivated by the exploratory drive. This natural intellectual work usually is not as complicated as rocket science and does not require taking college philosophy classes. The normal human intelligence is motivated to *ask questions* about the criteria for intrinsically valuing or holding oneself responsible for any given creature's well-being, and taking related actions. If we were to read a novel in which none of the characters ever ask themselves these kinds of questions, the reading would become tedious fairly quickly.

One of the thorniest objections we had to consider against this line of reasoning is that the sheer fact that *I value* X emotionally doesn't necessarily entail that I posit a truth-functional claim that X "*has* value." My approach to this question is based on the earlier clarification of the various meanings of "value." There is no coherent meaning to the phrase "X has value" without *some* relation to the idea of valuing beings valuing or dis-valuing things. As soon as we value anything, our basic emotions impel us already to be enmeshed in the realm of affirming the value of the well-being of at least some others qua valuing beings in their own right (family, close friends, etc.). When we realize that the criterion for intrinsically "valuing" a given creature doesn't depend on its instrumentality toward our own interests, we also see that the most coherent sense of "value" that can reasonably be used requires opening philosophical questions about the coherence of our criteria for directing this valuational concern in certain patterns as opposed to others.

Since the SEEKING system motivates the posing of these questions, yet does not prejudge their answers, whatever we end up believing is not necessarily pre-ordained by any combination of drive systems. The SEEKING system is therefore the crucial motivation to pursue truth-functional ethical conclusions rather than just believing whatever makes us feel good or whatever is habitual in our immediate social environment – even if it turns out that truth-functional statements can be grounded only in coherence epistemology rather than foundationalist deductive and/or empirical "proofs."

Given A. J. Ayer's demand for foundationalist "verificationism," an Ayer-style pure emotivist might try to defend the coherence of the view that "X has value" can mean *only* "I value X." It might be held that it is simply a *meaningless* or *trivial* statement if used in the other senses we have considered. We value things, but this doesn't entail that they "have" value. Therefore, there is no need to admit that what others value has any special status as something that I "ought" to help them to attain, even prima facie.

But in the first place, the argument that non-emotivist value statements are "meaningless" hinges on the "empiricist criteria for meaning," which in turn presuppose a foundationalist epistemology. It is foundationally impossible to "prove" the truth of an ethical statement, but foundational proofs are not appropriate for ethical statements. From the standpoint of the coherence epistemology for which I have argued, those who take this position will still find themselves questioning what the coherent criteria might be as to why there are *certain selected* others about whom we cannot avoid saying that their intrinsic value would warrant action on someone's part even if we ourselves had never existed.

Prior to any philosophical thinking, we naturally do extend the value of subjective valuational acts to some others beyond ourselves, and this is where the "problem of indexicals" becomes a factor in our coherence analysis. We are committed to thinking that at least in some cases a subject's valuational acts can create value simply because that subject values something – that these other subjects' values will matter even after we are dead or that their values were already important even before we came to know them personally.

Granted, if at least *some* sympathetic sentiments did not occur naturally, then there could be no motivation to be moral in a non-trivial sense, because there would be no ability to appreciate the value of others qua valuing creatures. More important, if we did not have the power to control our own attention, we would be unable to pay attention to the valuing dimension of others except in cases of hardwired sentiments. We would be like cats, who do intrinsically value the well-being of their own family members because of hard-wiring of their brains; but they are incapable of directing attention to the valuing dimension of mice or birds that they thoughtlessly torture while playfully killing them. Given that "ought implies can," there could be no epistemologically viable ethical statements without some natural empathic capacities of this kind, since there would be no natural motivations to trigger or even enable us to understand the meaning of moral thinking.

As soon as the exploratory drive kicks in, we are motivated to think seriously (with the cognitive tools available to our species) about the criteria for limiting or extending the realm of valuing creatures that count as having intrinsic value. At that point, we first have to contend with the "is/ought" problem and the danger of the naturalistic fallacy. We need for a moral worldview not just to tell us that we ought to do whatever we would have done anyway (whatever we were naturally motivated to do), but rather that there are certain ways to behave that are morally preferable to others, independently of what we happen to want. What is needed here is a resolution to the "ought implies can" problem in relation to moral motivation.

As we have seen, there is a remedy for this problem, but it requires establishing two important conclusions: first, the motivation for morality is not based just on feelings of altruism, empathy, or fellow feeling, but also on a motivation to discover what the truth actually is – i.e., the human version of the "exploratory drive." What is interesting about the exploratory drive that makes it different from "mere moral sentiments" is that the conclusions toward which it impels us are not predetermined by the sentiment itself. It drives us to want to know what the truth is about our world, *regardless* of whether the content of that truth is pleasant or unpleasant for us, or whether it is completely irrelevant to our well-being one way or the other. To the extent that this exploratory drive is not being suppressed, persuading people to think that certain moral statements are more or less likely to be true doesn't depend on their *wanting* those statements to be true or false; they want to know which statements *are actually* true, regardless of whether the content of the statements benefits them or not.

But secondly, we need to understand why anyone should be *motivated* to act in accordance with moral principles even if it is acknowledged that a coherent moral system includes the universalization of the value of valuing creatures. One of our primary concerns has been the question of the incoherence of value nihilism. If I say merely that I value things, but that a subject valuing something doesn't mean that it "has" value, this position is an endorsement of the thesis that the only meaning of the term "value" is subjective. It *defines* "value" as meaning nothing other than "I value X." It limits value to just what *one particular* subjective being values – myself. This position then has to contend with the coherence problems resulting from the problem of indexicals and the problem of natural altruistic feelings discussed above. The coherence issues become more complicated when we combine those two problems. The question is how we can index intrinsic (not merely instrumental) value only to certain specific individuals, when we are already committed to the idea that the sheer fact that they value something creates value, and would have created value even if we ourselves had never existed.

Because of the self-organizational and self-moving nature of human beings, we can't avoid valuing something. If we valued nothing, we would do nothing. The existence of non-consummatory drives, such as Frijda's "readiness potentials" and Panksepp's SEEKING system, has the important implication that the organism is the agent of its own actions rather than a passively reacting machine.

If active creatures are to act, then they must value something. As soon as we have posited that something has value, unless we are extreme sociopaths, we realize that the value of at least *certain* specific others is part of what moves us to act. A person with a fully functioning human brain will have CARE, SEPARATION DISTRESS, and in general social bonding emotion systems, which hardwire us to care about the subjective valuations of certain *specific* others in our lives. Those too would then have to be commensurable both with egoistic valuations and with any "ultimate," completely supra-subjective values we might believe to exist (if any). Our brains are hardwired to care – *intrinsically*, not just instrumentally – about the subjective valuations of at least certain narrowly selected others – mates, family members, etc. Dennett and Singer argue this on evolutionary grounds, and

Panksepp and other neuropsychologists show the neurophysiological basis of it.

But if we were to leave it there, we would still be trapped in the naturalistic fallacy, as we saw earlier. At best, if that much were all there is, then even extreme "moral saints" would be incapable of universalizing altruistic sentiments, let alone ethical mores and principles of justice. So we also have to consider what happens when the human version of the exploratory drive is added to the mix. In that case, we will search for the most coherent possible overall value system. It is only after we have been able to generalize an appreciation of the "what it's like" dimension not just of our friends but of everyone, including the enemy, that we can coherently pose *ethical* questions, such as questions of justice and rights, even as those pertain to the friend.

If I am an egoistic hedonist, I believe that my subjective valuing of X is sufficient to render the statement "I ought to attain X" true, at least prima facie. But the social bonding, nurturance, and sexuality systems also force us to extend the capacity of subjects to create value beyond ourselves. We are committed to the thesis that (at least some) others too can create subjective value by virtue of valuing things. The SEEKING system, if fully activated, then motivates us to pose the following question: Suppose my valuing something is enough to show that it has prima facie value, i.e., that I ought prima facie to attain it. Is there some reason to limit this implication – that something has value by virtue of a subject's valuing it – to one or a few subjects and not others? What is the real criterion for this limitation? What I have tried to show is that the most coherent answer to this question is that we acknowledge the prima facie value of any given valuing creature to the extent that we focus attention on their valuing dimension.

As we observe the relevant attention patterns, even in cases of involuntary attention, we can notice that we increasingly value a friend's well-being *intrinsically*, not just instrumentally; that the more we focus attention on a friend's "what it's like" quality, the more we feel that their well-being must already have had value with or without the addition of our feeling that it had value. The value that we posit for the friend's well-being is not contingent on our own existence at the time of the well-being that is valued. Our appreciation of the friend's intrinsic value correlates with our increasing attentional focus on the person's valuing dimension. So universalizing a moral sentiment is partly a matter of controlling our voluntary attention.

Psychologically, universalized compassion has a hypothetical status. It is something we always *could* direct attention toward if we were to choose. Like the back side of a house, the intrinsic value of those to whom we are not currently directing empathic attention is not something that we can believe doesn't exist. It is something that we know we *could* walk around and look at any time we choose. A key aspect of this "general" appreciation for the value of unknown creatures (even beyond the horizon of our immediate experience) depends on a phenomenological learning process. But we also make a series of inferences – not just deductive inferences but also coherence-based thought processes in the interest of avoiding an incoherent hermeneutical worldview. The *motivation* to undergo this learning process, as I have argued, is innate rather than socially constructed.

One of the coherence challenges faced by egoism or even the expanded egoism of "clannism" or "ethnocentricity" is to explain why it is that my subjectivity can give rise to value whereas other subjectivities, which are very similar to mine in the relevant respects, cannot. A still more formidable challenge is to explain why the supposed egoist allows that the valuations of certain selected others (for example, family members) are not contingent on the egoist's own valuations. How is it coherent to maintain that the value of my children's well-being will persist after I die, yet at the same time that the value of their well-being is contingent on my egoistically conferring that value upon them? This question in turn leads to all the other coherence problems we discussed in relation to failure to acknowledge the intrinsic value of all valuing creatures per se, at least prima facie.

When we think about the criteria that we apply in selecting *whose* well-being to intrinsically value, we can notice phenomenologically that the consistent criterion is the extent to which we have come to focus our attention on the valuing dimension of their "what it's like" quality. The criterion is that we recognize the value of valuing creatures to the extent that we choose to focus attention on them; thus all valuing creatures have value in the sense that if I *were* to focus much attention on them, I *would* have to recognize their value on pain of incoherent justifications for not doing so. This means that their valuations create value independently of whether I am currently focusing attention on this value or not.

Why should I say that a casual acquaintance's values have only minimal value, yet as the indexical Ralph Ellis gets to know that acquaintance better and better, the formerly valueless person becomes so valuable that I now take the position that this person's well-being will continue to matter even after I am dead? A patchwork of ad hoc hypotheses could perhaps preserve the coherence of such a position, but the more complicated such scenarios become, the more need there seems to be for a more coherent accounting. It is much more coherent simply to admit that the person's well-being had value in what we called "sense (3) value" all along:

> (3) "X has value" = "*Any* rational being with the capacity for empathy would value X *if attention were directed* to X, on pain of incoherence."

I merely came to appreciate this value by directing attention to it in certain ways.

The voluntary control of selective attention allows us to focus on the valuing dimension of others, or to direct attention away from it, often depending on whether we are more strongly driven by an exploratory drive to ask philosophical questions, or on the contrary by confabulatory tendencies such as authoritarianism and the ad hoc conspiracy theories of totalitarian movements or fanatical religious cults. Voluntary control of attention also enables us to direct attention away from the valuing dimension of most creatures at most times, while still understanding that we *could* direct attention toward any one of them at any given time, and therefore understand that they have prima facie value in our sense (3). The way we direct our attention can also affect the degree to which we feel *inspired to act* in accordance with the sense (3) value that we apperceive in this way.

Conversely, I have suggested that an implicit or partial value nihilism can sometimes motivate us to deliberately direct attention *away* from the valuing dimension of certain categories of valuing creatures, as reflected by a difficulty in appreciating the value of valuing creatures per se. So one of the main components of obsessive hatred – especially when the hatred is not motivated by any specific victimization by the hated person or group – may well be that in hatred we direct outward the underlying, unacknowledged partial nihilism. In this case, we are living "on the edge of nihilism." To be sure, nihilism doesn't always lead to hatred, but in many instances hatred can be the result of an underlying value nihilism. The classic expression of this kind of hatred is in the "sudden murder," which nihilates the intrinsic value not only of the victim, but also of the perpetrator, who takes no precautions not to be apprehended, and has nothing to gain from the crime except the expression of anger and hatred.

It would be difficult to think that the statement "I have value by virtue of being a valuing creature" is true if uttered from the perspective of this person that I now am, *and also from the perspective of a narrow group chosen by this person* (that small circle of friends and tribe members who naturally draw my intrinsic valuation), yet not true if uttered from different perspectives. Such a value system would have to posit that certain selected people's subjectively valuing something is enough to make it have value in some instances, but that this same fact would not cause the valuing creature to have value in other arbitrarily designated instances. The question then arises: why should *these* particular subjectivities have value by virtue of being valuing creatures, whereas other valuing creatures do not have value by virtue of this same capacity? If I were to morph into a slightly different person from the one I now am, would my ability to ground value by valuing things lose its force? As we have seen, this is not a *logically impossible* position to take, but it does seem to call for some accounting for the coherence of such a position.

As acknowledged in our earlier discussion of A. J. Ayer's emotivist position, this problem *could* be avoided if we were to hold the position that "I value X" is the *only* meaningful definition of "X has value." After all, even in our "morphing" thought experiment, no matter how many of my features change, even to the point that I am no longer "the same" person, I am at each point *an* individual, so at each point, it is logically possible to assert that only the person that "I" am (at that time) can ground value. But the crucial coherence problem with this position is the same as the above. If I have a normally functioning emotional brain, I can't claim not to intrinsically value the valuations of anyone other than myself.

Secondly, the justification for disallowing other definitions of "X has value" depends on Ayer's "empiricist criterion for meaning." But as we have already seen, the empiricist criterion for meaning presupposes a foundational epistemology, whereas ethical thinking requires a coherence epistemology.

Even if we grant, as I have tried to show, that a true ethical nihilism in the strict sense (i.e., the idea that nothing is worth doing) can't be coherently affirmed by a creature with a brain like ours, the coherence approach faces still another challenge that is almost as daunting. In attempting to work out the most coherent possible overall value system, it would be a daunting task to assess the number of ungrounded assumptions and ad hoc hypotheses needed for every possible value system, and their consistency with all relevant empirical facts and carefully clarified phenomenological facts, each of whose truth in turn would require its own coherence assessment. This seems to be a hopelessly complicated task.

The task becomes somewhat less complicated, however, if we can identify what the main sources of incoherence are for all different types of ethical viewpoints. This is where both the empirical and phenomenological psychology of the hermeneutical process itself can be most helpful. This was one of the main reasons we saw for the importance of emotion in ethical theory.

One of the strong "resistances" we face in trying to form a coherent moral viewpoint is the danger of falling into the naturalistic fallacy. To say we should do whatever we are naturally motivated to do provides no guidance and is morally vacuous. But the same problem also affects a more recent iteration – that moral feelings result from the "mirror neuron" system in the mammalian brain. Stein Bräten (2007, 2013) explores the behavioral implications of what can happen when the mirror neuron system and other empathy-orchestrating brain functions are suppressed, resulting in immoral or inhumane behavior. And he shows how just the opposite kind of behavior – extremely altruistic and selfless action – is equally normal and typical for humans. But as far as the naturalistic fallacy problem is concerned, whether we are dealing with fellow feeling that results from natural selection or with the effects of a mirror neuron system, we still need to deal with essentially the same problem: "You ought to act morally when and if you are naturally motivated to feel altruistic" (or cooperative, or some other natural feeling) is always in danger of reducing to "Always do whatever you want."

To be sure, it now seems apparent that a mirror neuron system is necessary if a person is to develop morally – necessary, but not sufficient. We need a mirror neuron system to empathize with the other's point of view; but empathy alone is not enough to motivate a coherent or even non–self-deceptive moral viewpoint – especially when the empathy can be turned on and off in the ways that Bräten describes. In the final analysis, Bräten's view on this point is similar to my own, because his analysis shows that the failure of empathy is just as natural as empathy itself.

We saw that there is a continuum rather than a sharp distinction between what people often call "moral" and "non-moral" values. All values have to do with deciding what we ought to do, whether the purpose is ultimately to achieve something for ourselves or others, or simply to enact a kind of action that is valuable in its own right – such as play, artwork, or eulogizing a loved one. The balance between our own well-being and others' is part of the ethical problematic as a whole. So it is crucial to explore the phenomenology of the way we actually come to believe that we "should" do this or that, whether in a "moral" or a "non-moral" sense. What we "should" do for our own purposes, whatever it might be, can't be categorized simply as "non-moral" value. Not only does my own well-being have the same value as anyone else's, but also I must weigh various values against each other, including the "moral" and "non-moral" ones, in order to decide what to do at any given moment. Whether I "should" eat at a French restaurant or donate the money to hungry children is simultaneously both a "moral" and a "non-moral" decision as to what I "should" do, all things considered. The "moral" and the "non-moral" must be commensurable for the purposes of decision. And this implies that the concept of value itself must be defined in relation to the valuing function of valuing creatures.

In terms of evaluating the coherence of overall belief systems, we have to remember that belief systems are formed and modified in a hermeneutical fashion. This means that each step in the thought process is motivated partly by a desire to keep intact the previous overall worldview and to focus attention predominantly on congruent facts and arguments. Only when a fact or argument can't easily and reasonably be fit into the overall worldview does the latter change. Primarily for this reason, we noted the necessity to explore in what ways poor phenomenological reflection on the meaning of our emotions is a major contributor – in my view, *the* major contributor – to incoherence in the lifelong project of developing a moral viewpoint.

The relationship between religion and morality entails a similar hermeneutical process. Our overall philosophical thinking always already plays a major role in determining what we believe about religion, because it largely determines the way we think it *makes the most sense* to interpret a text. The Christianity of Martin Luther King was not the Christianity of Alabama Governor George Wallace, whose famous proclamation "Segregation forever!" was echoed (in essence) by many preachers from their pulpits, who managed to interpret various biblical texts to support it. And this, again, means that philosophical thinking is more basic than religion in determining the ethical beliefs even of the most religious people, whether they are aware of it or not.

But we saw that logic alone is not an adequate tool for evaluating the comparative coherence of different moral views. Nor is the determination of which type of "social contract" is most legitimate easily based on purely foundational logical arguments. Can a logical argument, all by itself, prove whether or not the *retributive* principle of "giving people what they morally deserve" irrespective of the consequences (presumably based on their own character and actions) should or should not (contrary to Rawls) *supersede* the demands of distributive justice to allocate opportunities and resources on an equitable basis? Could either distributive or retributive justice be proven to take priority over the goal of maximizing utility, or vice versa? Could any or all of these principles be proven to take priority over our own self-interest in a particular situation?

If moral action is to be motivated, yet at the same time universalized, then it must involve the interplay of two different *types* of emotional systems in the human organism. One is already somewhat well understood – the altruistic emotions such as nurturance and social bonding which can lead to Hume's "fellow feeling" and Scheler's "sympathy." The other type of emotional system is just beginning to be understood and studied by psychologists, as we have seen in the previous chapters. The "exploratory drive" is present in mammals and most other higher animals, but doesn't lead to a search for universalized moral truth until the thinking capacity of humans is attained. In the process of reflecting on the subtle interplays of these two different types of emotional processes – natural sympathy on the one hand, and the "exploratory drive" on the other – it is crucial to use careful phenomenological methods.

In this regard, emotions make a strong contribution to the overall shape of a hermeneutical picture. In the process, it becomes increasingly obvious that interpreting our own emotions without serious distortions of this kind is a much more difficult process than many people suppose.

We have seen that the hermeneutic circle is crucially important for purposes of moral thinking because, contrary to most of the tradition of academic moral philosophy, the truth of ethical statements – just as with scientific statements – can't be absolutely "proven" foundationally. Epistemological and other implicit presuppositions of such proofs always already must be questioned. We always bring presuppositions *to* any experience, and this inevitable fact extends also to any attempt to "know ourselves" – hence the hermeneutic circle. We often "re-interpret" the new experience to make it fit the presuppositions rather than engaging in a potentially threatening upheaval of our entire pre-existing worldview. Even phenomenological reflection reveals that this is the way our minds work. Our preconceptions "propose," and reality "disposes" – but it disposes only of those realities toward which we direct our selective attention. And the way we direct our attention may in turn be predetermined by the preconceptions themselves.

We also considered the other main relevance of the psychology of emotion – the implications it has for logical problems in the coherence of various moral viewpoints for purely epistemological reasons. For example, we saw that the need for a motivation for moral action, combined with the "ought implies can" principle and the "is-ought" problem, has serious implications for which kind of ethical viewpoints can hold up under coherence-oriented criticism. Many moral notions have been developed as ways to avoid empathizing with others and to justify lack of attention to various ethical concerns, as well as to block out existentially troubling issues, as in the psychology of hermeneutics taken up in the theories of Arendt, Becker/Greenberg, and Adorno/Altemeyer that we discussed earlier.

In concrete phenomenological terms, the reason a previously unknown person can come to occupy the "inner circle" of people we care about qua specific individuals is that we direct more and more of our attention to the vicarious experiencing of what that person's valuational subjectivity "is like." Conversely, the way we *avoid* actually feeling the prima facie value of what some others posit as valuable is by directing our attention *away* from the "what it's like" dimension of that person's subjective needs or aspirations so that we don't empathize with them, as we do with mates, children, tribe members, etc. In other words, the way we avoid feeling this valuation is by avoiding any vicarious re-enactment of their valuational experience by directing our own attention away from that dimension.

We can't constantly empathize with the suffering of every individual conscious being in the world. But if we think through the implications of the naturalistic and phenomenological facts discussed earlier, we are virtually forced to admit that if we *did* focus attention on the hungry children in India as much as we do on our own children, we *would* feel the value of their subjective valuations. Universalized empathy has a conditional and counterfactual status in this respect. We know preconsciously that *if we were* to focus this much attention on any one of the millions of unknown hungry children, we would empathize with their valuations and correlative suffering. But even as a merely counterfactual awareness, these facts have to affect our thinking about what is the best way to distribute the effects of our actions.

The fact that the project of developing a coherent viewpoint resists our attempts suggests that there is a truth independent of what we might fancy it to be. Without resistance, we could believe whatever we want, but the demand for coherence causes many moral viewpoints to resist our attempts to believe them. Coherence in this sense requires more than merely rendering logically consistent a set of views initially grounded only in "moral intuitions." It requires acknowledging the difficulty of an inevitably hermeneutical process in which we already always have a pre-existing worldview in terms of which we tend to interpret new experiences. And part of the implication of this hermeneutical process is that our motivating emotions determine the direction of attention, and resulting belief systems. The main contributor to the coherence or incoherence of a belief system is therefore the murky realm of affective processes that play the most crucial role in directing the hermeneutical process, and resulting coherence of the system. Complicating the situation still further is the need for moral values to be felt with enough intensity to inspire us to action. Luckily, however, we are inherently acting creatures, and therefore cannot coherently endorse value nihilism.

References

Adorno, T., E. Frenkel-Brunswik, D. Levinson, and R. N. Sanford (1964). *The Authoritarian Personality*. New York: Wiley.

Almeder, R. (1986). Fallibilism, coherence and realism. *Synthese* 68: 213–223.

Altemeyer, R. (2008). *The Authoritarians*. Ramona, CA: Cherry Hill Publishing.

Anderson, J. L., J. Wisnowski, H. Barrash, H. Damasio, and D. Tranel (2009). Consistency of neuropsychological outcome following damage to prefrontal cortex in the first years of life. *Nature Neuroscience* 31: 170–179.

Arendt, H. (1959). *The Human Condition*. Chicago, IL: University of Chicago Press.

(1963). *Eichmann in Jerusalem: a Report on the Banality of Evil*. London: Faber & Faber.

(1968). *The Origins of Totalitarianism*. New York: Harcourt/Harvest.

Aristotle (1966). *Nicomachean Ethics*. New York: Bobbs-Merrill.

Aroles, S. (2007). *L'énigme des enfants loup*. Paris: Éditions publibook.

Auxier, R. (1995), The wind we inherited: God and secular America, *Personalist Forum*, 112: 95–124.

Ayer, A. J. (1946). *Language, Truth and Logic*. London: Gollancz.

(1965). On the analysis of moral judgments. In *Philosophical Essays*. New York: St. Martin's Press.

Barbas, H. and B. Zikopoulos (2006). Sequential and parallel circuits for emotional processing in primate orbitofrontal cortex. D. H. Zald and S. L. Rauch (eds.) *The Orbitofrontal Cortex*. Oxford: Oxford University Press, pp. 57–94.

Becker, E. (1973). *The Denial of Death*. New York: The Free Press.

Bell, C. (1978). Racism, narcissism, and integrity. *Journal of the National Medical Association* 70: 89–92.

(1980). Racism: A symptom of narcissistic personality disorder. *Journal of the National Medical Association* 72: 661–665.

Ben Ze'ev, A. (2000). *The Study of Emotions*. Boston, MA: MIT Press.

(2002). Emotions are not feelings. *Consciousness & Emotion* 3(1), 81–89.

Bettelheim, B. (1959). Feral children and autistic children. *American Journal of Sociology* 64, pp. 455–467.

Binswanger, L. (1963). *Being in the World*. New York: Basic Books.

Borosini, V. (1913). Problem of illegitimacy in Europe. *Journal of Criminal Law and Criminology* 4: 212–236.

Bowles, S., and H. Gintis (2013). *A Cooperative Species: Human Reciprocity and Its Evolution*. Princeton, NJ: Princeton University Press.

Brand, G. (1967). Intentionality, reduction, and intentional analysis in Husserl's later manuscripts. In J Kockelmans (ed.), *Phenomenology*. Garden City, NY: Doubleday, pp. 197–217.

Bräten, S. (ed.) (2007). *On Being Moved: From Mirror Neurons to Empathy*. Amsterdam: John Benjamins.

(2013). *Roots and Collapse of Empathy*. Amsterdam: John Benjamins.

Briere, J. B., B. A. Nelson, and P. A. Krol (1987). Childhood sexual and physical abuse as factors in adult psychiatric illness. *American Journal of Psychiatry* 144: 1426–1430.

Brightman, E. S. (1933). *Moral Laws*. New York: Abingdon Press.

Buford, T. (2011). *Self-Knowledge, An Essay in Social Personalism*. Lanham, MD: Lexington Books.

Bühler, C. (1931). *Kindheit und Jugend*. Leipzig: Hirzel.

Bush, G., P. Luu, and M. I. Posner (2000). Cognitive and emotional influences in anterior cingulate cortex. *Trends in Cognitive Sciences* 4 (6): 215–222.

Camus, A. (1963). *The Fall*. New York: Vintage.

Chareyron, L. J., P. B. Lavenex, D. G. Amarel, and P. Lavenax (2012). Postnatal development of the amygdala: A stereological study in macaque monkeys. *Journal of Comparative Neurology*. June 15: 520(9): 1965–1984.

Cook, S. (1939). The production of experimental neurosis in the white rat. *Psychosomatic Medicine* 1: 2, 293–308.

Cornell, D. (1995). *At the Heart of Freedom*. Princeton, NJ: Princeton University Press.

Cross, T. (1987). *The Black Power Imperative (1987)*. Gainesville, FL: Faulkner Press.

Damasio, A. (1999). *The Feeling of What Happens*. New York: Harcourt Brace.

(2003). *Looking for Spinoza*. (New York: Harcourt).

Daniels, N. (1979). Wide reflective equilibrium and theory acceptance in ethics. *The Journal of Philosophy* 76: 256–282.

Danto, A. C. (1968/1971). *What Philosophy Is*. New York: Penguin.

Dawkins, R. (1976/2006). *The Selfish Gene*. Oxford: Oxford University Press.

De Beauvoir, S. (1948/1976). *The Ethics of Ambiguity*. New York: Kensington Press/Philosophical Library.

Dennett, D. (1978/2001). Where Am I? In D. Hoffstadter and D. Dennett (eds.), *The Mind's I*. New York: Basic Books.

(1984). *Elbow Room: The Varieties of Free Will Worth Wanting*. Boston, MA: MIT/Bradford.

(2003). *Freedom Evolves*. New York: Viking.

DePaul, M. (1986). Reflective equilibrium and foundationalism. *American Philosophical Quarterly* 23: 59–69.

De Tocqueville, A. (1835/2000). *Democracy in America*. Chicago: University of Chicago Press.

De Waal, F. (2006). *Primates and Philosophers: How Morality Evolved.* Princeton, NJ: Princeton University Press.

(2008). Putting the altruism back into altruism: The evolution of empathy. *Annual Review of Psychology* 59: 279–300.

Descartes, R. (1649/1989). *The Passions of the Soul.* New York: Hackett.

Dieter, R. (1995). On the front line: Law enforcement views on the death penalty. Death Penalty Information Center, 1015 18th St., Washington, DC 20036, www.ncjrs.gov/App/Publications/abstract.aspx?ID=153848 (accessed August 23, 2017).

Duncker, K. (1939). Ethical relativity? An enquiry into the psychology of ethics. *Mind* (Oxford University Press) 48 (189): 39–57.

Dunlop, D. W., and C. B. Demeroff (2007). The role of dopamine in the pathophysiology of depression. *Archives of General Psychiatry* 64: 327–337.

Duriez, B., and B. Soenens (2009). The intergenerational transmission of racism: The role of right-wing authoritarianism and social dominance orientation. *Journal of Research in Personality* 43, 906–909.

Ellis, R. D. (1985). *An Ontology of Consciousness.* Berlin: Springer.

(1992). *Coherence and Verification in Ethics.* Washington, DC: University Press of America.

(1995). *Curious Emotions.* Amsterdam: John Benjamins.

(1996). *Eros in a Narcissistic Culture.* Berlin: Springer.

(1998). *Just Results: Ethical Foundations for Policy Analysis.* Washington, DC: Georgetown University Press.

(2004). *Love and the Abyss: An Essay on Value and Finitude.* Chicago, IL: Open Court.

(2005). *Curious Emotions: Roots of Consciousness and Personality in Motivated Action.* Amsterdam: John Benjamins.

(2008a). In what sense is 'rationality' a criterion for emotional self-awareness? *Consciousness and Cognition* 17: 972–973.

(2008b). The phenomenology of alexithymia as a clue to the intentionality of emotion. In L. Charland and P. Zachar (eds.), *Fact and Value in Emotion.* Amsterdam: John Benjamins, pp. 181–192.

(2013). Neuroscience as a human science. *Human Studies* 36: 491–507.

Ellis, R. D., and N. Newton (2010). *How the Mind Uses the Brain.* Chicago, IL: Open Court.

Emory University Health Sciences Center (2006). Emory study lights up the political brain. *Science Daily*, January 31. www.sciencedaily.com/releases/2006/01/060131092225.htm (accessed August 3, 2017).

Eshun, S., and R. Gurung (2009). *Culture and Mental Health: Sociocultural Influences, Theory, and Practice.* New York: Wiley-Blackwell.

Faw, W. (2000a). My amygdala-orbitofrontal circuit made me do it. *Consciousness & Emotion* 1: 167–179.

(2000b). Consciousness, motivation, and emotion: Biopsychological reflections. In R. D. Ellis and N. Newton (eds.), *The Caldron of Consciousness: Motivation, Affect, and Self-organization.* Amsterdam: John Benjamins.

(2003). Pre-frontal executive committee for perception, working memory, attention, long-term memory, motor control, and thinking: A tutorial review. *Consciousness and Cognition* 12: 83–139.

(2015). Neuropsychology in childhood and adolescent development. Lecture at *MidSouth Philosophy Conference.*

Federal Bureau of Investigation (1964). Counterintelligence Program Internal Security, Disruption of Hate Groups, National Knights of the Ku Klux Klan (Freedom of Information and Privacy Acts). https://vault.fbi.gov/cointelpro/White%20Hate%20Groups/cointelpro-white-hate-groups-part-04-of-14 (accessed August 3, 2017).

Festinger, L. (1957). *A Theory of Cognitive Dissonance.* Redwood City, CA: Stanford University Press.

Filonowicz, J. D. (2008) *Fellow-Feeling and the Moral Life.* Cambridge: Cambridge University Press.

Foucault, M. (1994). *The Order of Things: An Archeology of the Human Sciences.* New York: Vintage.

Fox, M. D., R. L. Buckner, M. P. White, M. D. Greicius, and A. Pascual-Leone (2012). Efficacy of transcranial magnetic stimulation targets for depression is related to intrinsic functional connectivity with the subgenual cingulate. *Biological Psychiatry* 72: 595–603.

Freud, S. (1925/1959). *Beyond the Pleasure Principle.* New York: Bantam.

Frijda, N. H. (2006/2007). *The Laws of Emotion.* London: Erlbaum.

Gabbay, V., B. A. Ely, Q. Li, S. D. Bangaru, A. M. Panzer, C. M. Alonso, F. X. Castellanos, and M. P. Milham (2013). Striatum-based circuitry of adolescent depression and anhedonia. *Journal of the American Academy of Childhood and Adolescent Psychiatry* 52: 628–641.

Gadamer, H. G. (1971/1975). *Truth and Method.* New York: Seabury.

Gallagher, S. (2006). *How the Body Shapes the Mind.* Oxford: Clarendon Press.

Gallese, V. Fadiga, L. Fogassi, and G. Rizzolatti (1996). Action recognition in the premotor cortex. *Brain* 119: 593–609.

Gee, D. G., K. L. Humphreys, J. Flannery, B. Goff, E. H. Telzer, M. Shapiro, T. A. Hare, S. Y. Bookheimer, and N. Tottenham (2013). A developmental shift from positive to negative connectivity in human amygdala – prefrontal circuitry. *Journal of Neuroscience.* March 6, 2013, 33(10): 4585–4593.

Gendlin, E. (1978/1982/1990). *Focusing.* Toronto: Bantam.

(1992). Thinking beyond patterns: Body, language, and situations. In B. den Ouden and M. Moen (eds.), *The Presence of Feeling in Thought* (New York: Peter Lang).

(1962/1997). *Experiencing and the Creation of Meaning.* Toronto: Collier-Macmillan/Chicago, IL: University of Chicago Press.

Gewirth, A. (1980). *Reason and Morality.* Chicago, IL: University of Chicago Press.

Gibson, J. J. (1986). *The Ecological Approach to Visual Perception.* New York: Routledge/Psychology Press.

Gilligan, C. (1982). *In a Different Voice: Psychological Theory and Women's Development.* Cambridge, MA: Harvard University Press.

Giorgi, A. (1971). Phenomenology and experimental psychology. In A. Giorgi (ed.), *Duquesne Studies in Phenomenological Psychology, Vol. I*. Pittsburgh, PA: Duquesne University Press, pp. 6–29.
Goleman, D. (1995/2005). *Emotional Intelligence*. New York: Bantam.
(1996). *Vital Lies, Simple Truths: The Psychology of Self Deception*. New York: Simon & Schuster.
Greenberg, J., S. Solomon, and T. Pyszczynski (1997). Terror management theory of self-esteem and cultural worldviews: Empirical assessments. *Advances in Experimental Social Psychology* 29 (S 61): 139.
Guardian, The (2012). "Legitimate rape" rarely leads to pregnancy, claims US Senate candidate. August 19, 2012. www.theguardian.com/world/2012/aug/19/republican-todd-akin-rape-pregnancy (accessed August 3, 2017).
Guthrie, R. D. (2005) *The Nature of Paleolithic Art*. Chicago, IL: University of Chicago Press.
Harlow, H. F. (1950). Learning motivated by a manipulation drive. *Journal of Experimental Psychology* 40: 228–234.
Harmon, G. (1986). Moral explanations of moral facts: Can moral claims be tested against moral reality? *Southern Journal of Philosophy* 24: 57–68.
Harmon, R., R. Gaensbauer, and T. Emde (1976). *Emotional Expression in Infancy: A Biobehavioral Study*. New York: International Universities Press.
Harvey, C. (2017). Human(un)kind and the rape of the world. *Philosophy in the Contemporary World* 23 (forthcoming).
Harvey, G. (2005). *Animism: Respecting the Living World*. New York: Columbia University Press.
Heidegger, M. (1927/1962). *Being and Time*. New York: Harper & Row.
Hentoff, N. (1973). *Five Ways of Making It*. New York: Viking Press.
Herman, J. (1992/1997). *Trauma and Recovery: The Aftermath of Violence – from Domestic Abuse to Political Terror*. New York: Basic Books.
Hibbing, J. R., K. Smith, and J. Alford (2013). *Predisposed: Liberals, Conservatives, and the Biology of Political Differences*. New York: Routledge.
Hicks, B., R. F. Kreuger, W. Iacome, M. McGue, and C. J. Patrick (2004). Family transmission and heritability of externalizing disorders: a twin-family study. *Archives of General Psychiatry* 61 (9): 922–928.
Hodson, G., S. Hogg, and C. McInnis (2009). The role of "dark personalities" (narcissism, Machiavellianism, psychopathy), Big Five personality factors, and ideology in explaining prejudice. *Journal of Research in Personality* 43: 686–690.
Hountondji, P. (1987). *African Philosophy*. Berlin: Springer.
Hsee, C. K., and B. Ruan (2016). The pandora effect: The power and peril of curiosity. *Psychological Science* 10, 1–8. See also: www.sciencedaily.com/releases/2016/04/160408101933.htm (accessed August 3, 2017).
Huffington Post (2016). Huffpolster; Most Republicans still won't say Obama was born in the U.S. *Huffington Post*, September 21, 2016. www.huffingtonpost.com/entry/trump-birther-poll_us_57e27935e4b0e80b1b9f30c0 (accessed August 3, 2017).

Hull, C. (1943/1966). *Principles of Behavior.* New York: Appleton-Century-Crofts.

Hume, D. (1740/1955). *A Treatise of Human Nature.* New York: Bobbs-Merrill.

Husserl, E. (1900/2001) *Logical Investigations.* New York: Humanity Books.

(1913/1931). *Ideas* (W. R. Boyce Gibson trans.). London: Collier 1931; from *Ideen zu einer Reinen Phänomenologie und Phänomenologischen Philosophie,* 1913.

(1936/1970). *The Crisis of European Sciences and Transcendental Philosophy* (David Carr trans.). Evanston, IL: Northwestern University Press.

Jeannerod, M. (1994). The representing brain: Neural correlates of motor intention and imagery. *Behavioral and Brain Sciences,* 17, 187–244.

(1997). *The Cognitive Neuroscience of Action.* Oxford: Blackwell.

Jennings, P. (1948). Report on resistentialism. *The Spectator,* April 23.

Jonas, H. (1966/2001). *The Phenomenon of Life: Toward a Philosophical Biology.* Evanston, IL: Northwestern University Press.

Joyce, R. (2008). What neuroscience can (and cannot) contribute to metaethics. In W. Sinnott-Armstrong (ed.), *Moral Psychology, Vol. 3: The Neuroscience of Morality.* Boston, MA: MIT Press, pp. 371–594.

Kagan, J., and M. Berkun (1954). The reward value of running activity. *Journal of Comparative Physiological Psychology* 47: 108–110.

Kant, I. (1781/1996). *Critique of Pure Reason.* Indianapolis, IN: Hackett.

(1788/2002). *Critique of Practical Reason.* New York: Hackett.

(1797/1996). *The Metaphysics of Morals.* Cambridge: Cambridge University Press.

Kauffman, S. (1993). *The Origins of Order.* Oxford: Oxford University Press.

Keeley, L. (1997) *War before Civilization: The Myth of the Peaceful Savage.* Oxford: Oxford University Press.

Kelly, R. C. (2000) *Warless Societies and the Origin of War.* Ann Arbor, MI: University of Michigan Press.

Kim, E., E. C. Lauterbach, A. Reeve, D. B. Arciniegas, K. L. Coburn, M. F. Mendez, T. A. Rummans, and E. C. Coffey (2007). Neuropsychiatric complications of traumatic brain injury: a critical review of the literature. *Journal of Neuropsychiatry and Clinical Neuroscience* 19: 106–127.

Kim, S. H., and S. Hamann. (2007). Neural correlates of positive and negative emotion regulation. *Journal of Cognitive Neuroscience* 19: 776–798.

Klein, M. H., P. L. Mathieu, E. T. Gendlin, and D. J. Kiesler (1969). *The Experiencing Scale: A Research and Training Manuel, Vol. 1.* Madison, WI: Wisconsin Psychiatric Institute.

Koenigs, M. and D. Tranel (2006). Pseudopsychopathology: a perspective from cognitive neuroscience, In D. H. Zald and S. L. Rauch (eds.), *The Orbitofrontal Cortex.* Oxford: Oxford University Press, pp. 597–620.

Kohlberg, L. (1973). The claim to moral adequacy of a highest stage of moral judgment. *The Journal of Philosophy* 70 (18): 630–646.

(1981). *Essays on Moral Development, Vol. I: The Philosophy of Moral Development.* San Francisco, CA: Harper & Row.

(1984). *The Psychology of Moral Development: The Nature and Validity of Moral Stages (Essays on Moral Development, Volume 2)*. San Francisco, CA: Harper & Row.
Kohut, H. (1985). *Self Psychology and the Humanities*. New York: W. W. Norton.
Kramer, M. H., and N. E. Simmonds (1996). Reason without reasons: A critique of Alan Gewirth's moral philosophy. *The Southern Journal of Philosophy* 34: 301–315.
Kuhn, T. (1964/2012). *The Structure of Scientific Revolutions*. Chicago: University of Chicago Press.
Lakoff, G. (1996). *Moral Politics*. Chicago, IL: University of Chicago Press.
(2008). *The Political Mind: Why You Can't Understand 21st Century Politics with an 18th Century Brain*. New York: Penguin/Viking.
Landgrebe, L. (1966). *Major Problems in Contemporary Philosophy*. New York: Ungar.
Langness, L. L. (1977). *Other Fields, Other Grasshoppers*. Philadelphia, PA: J. B. Lippincott.
Levinas, E. (1969). *Totality and Infinity*. The Hague: Martinus Nijhoff.
Locke, J. (1688/1989). *The Two Treatises of Civil Government*. London: Routledge.
Luce, P. A. (1960). *The White Citizens Councils of Mississippi*. Master's Thesis, Ohio State University.
Lungu, O., T. Liu, T. Waechter, D. T. Willingham, and J. Ashe (2007). Strategic modulation of cognitive control. *Journal of Cognitive Neuroscience* 19: 1302–1315.
Luria, A. R. (1980). *Higher Cortical Functions in Man*. New York: Basic Books.
Lyotard, J.-F. (1984). *The Postmodern Condition: A Report on Knowledge*. Minneapolis, MN: University of Minnesota Press.
MacIntyre, A. (1984). *After Virtue*. Notre Dame, IN: University of Notre Dame Press.
(1989). *Against the Self-Images of the Age*. Notre Dame, IN: University of Notre Dame Press.
Maslow, A. (1962/1970). *Motivation and Personality*. New York: Harper & Row.
May, R. (1950/2011). *The Meaning of Anxiety*. New York: Martino.
(1969). *Love and Will*. New York: W. W. Norton.
Mayberg, H. S., A. M. Lozano, and V. Voon (2005). Deep brain stimulation for treatment-resistant depression. *Neuron* 45: 651–660.
Merleau-Ponty, M. (1945/1962). *Phenomenology of perception* (C. Smith trans.). New York, Humanities Press. Original French edition 1945.
(1942/1963) *The Structure of Behavior* (A. Fischer trans.). Boston, MA: Beacon. Original French edition 1942.
Milgram, S. (1974). *Obedience to Authority: An Experimental View*. New York: Harper-Collins.
Mill, J. S. (1863/2015). *Utilitarianism*. London: Parker, Son & Bourn.
(1859/1947). *On Liberty*. New York: Appleton-Century-Crofts.
Montgomery, K. G. (1954). The role of the exploratory drive in learning. *Journal of Comparative Physiological Psychology* 47: 60–64.

Moore, G. E. (1900) *Ethics*. Oxford: Oxford University Press.
Moti, N. (1990). A cognitive reinterpretation of Stanley Milgram's observations on obedience to authority. *American Psychologisty* 46: 1384–1385.
Narveson, J. (1980). Gewirth's Reason and Morality – a study in the hazards of universalizability in ethics. *Dialogue* 19: 652–674.
NBC New York (2016). NYPD releases suspect sketch after attack on Muslim MTA worker. December 7, 2016. www.nbcnewyork.com/news/local/police-nypd-release-suspect-sketch-terrorist-grand-central-muslim-assault-attack-mta-employee-405108116.html (accessed August 3, 2017).
Neisser, U. (1976). *Cognition and Reality*. San Francisco, CA: Freeman.
New York Times (1992). After 350 years, Vatican says Galileo was right: It moves. October 31, 1992.
(1993). James R. Venable, 92 leader of Klan group. January 21 Obituaries.
(2012). Transcript of Bill Clinton's speech to the Democratic National Convention. September 5, 2012. www.nytimes.com/2012/09/05/us/politics/transcript-of-bill-clintons-speech-to-the-democratic-national-convention.html (accessed August 3, 2017).
News Herald, The (2013). Southerland: SNAP measure critics seek 'scapegoat'. June 29, 2013. http://mcdonaldworks.blogspot.com/2013/06/southerland-snap-measure-critics-seek.html#!/2013/06/southerland-snap-measure-critics-seek.html (accessed August 23, 2017).
Newton, N. (1996). *Foundations of Understanding*. Amsterdam: John Benjamins.
(2000). Conscious emotion in a dynamic system: How I can know how I feel. In R. Ellis and N. Newton (eds.), *The Caldron of Consciousness: Motivation, Affect, and Self-organization*. Amsterdam: John Benjamins, pp. 91–108.
Nielsen, K. (1989). *Why Be Moral?* New York: Prometheus.
O'Doherty, J. P. and R. J. Dolan (2006). The role of human orbitofrontal cortex in reward prediction and behavioral choice: insights from neuroimaging. In D. H. Zald and S. L. Rauch (eds.) *The Orbitofrontal Cortex*. Oxford: Oxford University Press, pp. 265–284.
Olds, J. (1956). Pleasure centers in the brain. *Scientific American*, October: 105–116.
Olson, S. (2002). *Mapping Human History*. Boston, MA/New York: Mariner/Houghton Mifflin.
Panksepp, J. (1998). *Affective Neuroscience*. New York: Oxford University Press.
(2000). The neuro-evolutionary cusp between emotions and cognitions: Implications for understanding consciousness and the emergence of a unified mind science. *Consciousness & Emotion* 1: 17–56.
(2011). Cross-species affective neuroscience decoding of the primal affective experiences of humans and related animals. *PloS (Public Library of Science) ONE* 6(8): e21236. http://journals.plos.org/plosone/article?id=10.1371/journal.pone.0021236 (accessed August 3, 2017).
Panksepp, J., J. Wright, M. Döbrössy, T. Schlaepfer, and V. Coenen (2014). Affective neuroscience strategies for understanding and treating depression: from preclinical models to three novel therapeutics. *Clinical Psychological Science* 2(4): 472–494.

Pavlov, I. P., and E. Gley (1929). *Leçons sur l'activité du cortex cérébral*. Paris: A. Legrand.

Pew Forum on Religion and Public Life (2015). U.S. Religious Landscape Survey. www.pewforum.org/2015/05/12/americas-changing-religious-landscape/ (accessed August 23, 2017).

Pojman, L., and J. Feiser (2005/2008). *Ethics: Discovering Right and Wrong*. Boston, MA: Wadsworth, Cengage Learning.

Posner, M. I. (1990). Hierarchical distributed networks in the neuropsychology of selective attention. In A. Caramazza, (ed.), *Cognitive Neuropsychology and Neurolinguistics: Advances in Models of Cognitive Function and Impairment*. New York: Plenum, pp. 187–210.

Prasad, M., A. J. Perrin, K. Bezila, S. G. Hoffman, K. Kindleberger, K. Manturuk, and A. Smith Powers (2009). "There must be a reason": Osama, Saddam, and inferred justification. *Sociological Inquiry* 79: 142–162.

Prinz, J. (2011). *The Emotional Construction of Morals*. Oxford: Oxford University Press.

Ragsdale, E. (2006). Value and meaning in gestalt psychology. In K. H. Dockett, G. R. Dudley-Grant, and C. P. Bankart (eds.), *Psychology and Buddhism*. Berlin: Springer, pp. 71–105.

Ralston, H. (1986). *Philosophy Gone Wild*. New York: Prometheus Books.

Rank, O. (1936/1978). *Truth and Reality*. New York: W. W. Norton.

Rankin, K. (2007). Social cognition in frontal injury. In B. L. Miller and J. L. Cummings (eds.), *The Human Frontal Lobes: Functions and Disorders*. New York: Guilford.

Rawls, J. (1971/1999). *A Theory of Justice*. Cambridge, MA: Harvard University Press.

(1993). *Political Liberalism*. New York: Columbia University Press.

Regis, E. (1982). *Gewirth's Ethical Rationalism*. Chicago, IL: University of Chicago Press.

Rescher, N. (1973) *Coherence Theory of Truth*. Oxford: University of Oxford Press.

Roberts, R. (2013). *Emotions in the Moral Life*. Cambridge: Cambridge University Press.

Rogers, C. (1951). *Client Centered Therapy: Its Current Practice, Implications, and Theory*. London: Constable.

(1961). *On Becoming a Person*. New York: Houghton Mifflin.

Ross, W. D. (1930). *The Right and the Good*. Oxford: Oxford University Press.

Roth, J. (1978). Prosecutor perceptions of crime seriousness. *Journal of Criminal Law and Criminology* 69: 232–242.

Rousseau, J.-J. (1762/1968). *The Social Contract*. New York: Penguin.

Salmela, M. (2014). *True Emotions*. Amsterdam: John Benjamins.

Sartre, J.-P. (1943/1965). *Being and Nothingness: An Essay on Phenomenological Ontology*, Trans. Hazel Barnes. New York: Washington Square Press.

Sayre-McCord, G. (1985). Coherence and models for moral theorizing. *Pacific Philosophical Quarterly* 6: 170–190.

Scheler, M. (1954/1970). *The Nature of Sympathy*. Hamden, CT: Archon Books.

Schleiermacher, F. (1838/1998). *Hermeneutics and Criticism: And Other Writings*. Cambridge: Cambridge University Press.

Schopenhauer, A. (1818/1966). *The World as Will and Representation*. New York: Dover.

Sellin, T., and M. Wolfgang (1964). *The Measurement of Delinquency*. Washington, DC: Institute for Law and Social Research.

Shemmer, M. (2010). The political brain. *Scientific American*, July 26.

Sher, G. (1987) *Desert*. Princeton, NJ: Princeton University Press.

Singer, P. (1981/2011). *The Expanding Circle*. Princeton, NJ: Princeton University Press.

Sinnott-Armstrong, W. (ed.) (2008), *Moral Psychology, Vol. 3: The Neuroscience of Morality*. Boston, MA: MIT Press, pp. 371–594.

Smart, J. J. C. (1984). *Ethics, Persuasion and Truth*. London: Routledge.

Sober, E., and D. S. Wilson (1998). *Unto Others: The Evolution and Psychology of Unselfish Behavior*. Cambridge, MA: Harvard University Press.

Stamenov, M. and V. Gallese (eds.) (2002). *Mirror Neurons and the Evolution of Brain and Language*. Amsterdam: John Benjamins.

Stein, E. (1916/1989). *The Problem of Empathy*. Washington, DC: ICS Publications.

Sterba, J. (1980). *The Demands of Justice*. Notre Dame, IN: University of Notre Dame Press.

Stout, M. (2005). *The Sociopath Next Door: The Ruthless versus the Rest of Us*. New York: Broadway Books.

Strasser, S. (1969). *The Idea of Dialogal Phenomenology*. Pittsburgh: Duquesne University Press.

Sussman, R., and R. Cloninger (2011). *Origins of Altruism and Cooperation*. Berlin: Springer.

Thornhill, N. (1993). *The Natural History of Inbreeding and Outbreeding: Theoretical and Empirical Perspectives*. Chicago, IL: University of Chicago Press.

Tottenham, N., T. A. Hare, and B. J. Casey (2009). Developmental perspective on human amygdala function. In P. J. Whalen and E. A. Phelps (eds.), *The Human Amygdala*. New York: Guilford Press, pp. 107–117.

Tremblay, L. K., C. A. Naranja, S. J. Graham, N. Hermann, H. S. Mayberg, S. Hevenor, and U. E. Busta (2005). Functional neuroanatomical substrates of altered reward processing in major depressive disorder revealed by a dopaminergic probe. *Archives of General Psychiatry* 62(11): 1228–1236.

Varela, F., E. Thompson, and E. Rosch (1991/1993). *The Embodied Mind*. Cambridge Mass.: MIT Press.

Watt, D. (1998). Affect and the 'hard problem': neurodevelopmental and corticolimbic network issues. *Consciousness Research Abstracts: Toward a Science of Consciousness, Tucson 1998*: 91–92.

(2000). The centrencephalon and thalamocortical integration: Neglected contributions of periaqueductal gray. *Consciousness & Emotion 1*: 91–114.

Wegner, D. (2003). *The Illusion of Conscious Will*. Cambridge, Mass.: MIT/Bradford.

Westen, D. (2008). *The Political Brain*. New York: Public Affairs.

White, R. W. (1959). Motivation reconsidered: the concept of competence. *Psychological Review* 66: 297.

Wilson, E. O. (1975/2000). *Sociobiology: The New Synthesis.* New York; Belknap Press.

Wohlschläger, A. and H. Bekkering (2002). The role of objects in imitation. In M. Stamenov and V. Gallese (eds.), *Mirror Neurons and the Evolution of Brain and Language.* Amsterdam: John Benjamins, pp. 101–114.

Yehuda, R. (1995). Impact of cumulative lifetime trauma and recent stress on current posttraumatic stress disorder symptoms in holocaust survivors. *American Journal of Psychiatry* 152: 1815–1818.

Yehuda, R., S. Halligan, and R. Grossman. (2001). Childhood trauma and risk for PTSD. *Development and Psychopathology* 13: 733–753.

Zachar, P. (2000a). *Psychological Concepts and Biological Psychiatry.* Amsterdam/Philadelphia: John Benjamins.

(2000b). Child development and the regulation of affect and cognition in consciousness: A view from object relations theory. In R. Ellis and N. Newton (eds.), *The Caldron of Consciousness: Motivation, Affect, and Self-organization.* Amsterdam: John Benjamins, pp. 205–222.

Zito, S. (2016). Taking Trump seriously, not literally. *The Atlantic*, September 23.

Index